HETEROGENEOUS PROCESSES OF GEOCHEMICAL MIGRATION

GETEROGENNYE PROTSESSY GEOKHIMICHESKOI MIGRATSII

ГЕТЕРОГЕННЫЕ ПРОЦЕССЫ ГЕОХИМИЧЕСКОЙ МИГРАЦИИ

HETEROGENEOUS PROCESSES
OF
GEOCHEMICAL MIGRATION

V. S. Golubev and A. A. Garibyants

Translated from Russian by
J. Paul Fitzsimmons
Professor of Geology
University of New Mexico
Albuquerque, New Mexico

 CONSULTANTS BUREAU · NEW YORK-LONDON · 1971

The original Russian text, published by Nedra Press in Moscow in 1968, has been corrected by the authors for the present edition. The English translation is published under an agreement with Mezhdunarodnaya Kniga, the Soviet book export agency.

Владимир Степанович Голубев
Артемий Арташесович Гарибянц

ГЕТЕРОГЕННЫЕ ПРОЦЕССЫ ГЕОХИМИЧЕСКОЙ МИГРАЦИИ

Library of Congress Catalog Card Number 73-140829

ISBN-13: 978-1-4684-1589-6   e-ISBN-13: 978-1-4684-1587-2

DOI: 10.1007/978-1-4684-1587-2

© 1971 Consultants Bureau, New York

Softcover reprint of the hardcover 1st edition 1971

A Division of Plenum Publishing Corporation
227 West 17th Street, New York, N.Y. 10011

United Kingdom edition published by Consultants Bureau, London
A Division of Plenum Publishing Company, Ltd.
Davis House (4th Floor), 8 Scrubs Lane, Harlesden, NW10 6SE, London, England

CONTENTS

Introduction 1

Chapter 1. Formulating the Problem of the Geochemical Migration of Included
      Substances, and Methods of Solving It 5
  § 1. Formulating the problem of geochemical migration of included sub-
     stances.. 5
  § 2. Equations of material balance and kinetics of the processes of inter-
     action between a substance and host rocks 5
  § 3. Hydrodynamic equations 8
  § 4. Simplification and methods of solving the problem 9
  § 5. The geochemical migration of mixtures 11
Literature cited 11

Chapter 2. Diffusion in Rocks 13
  § 6. Laws of diffusion 13
  § 7. Solutions of diffusion equations for steady-state current 14
  § 8. Solutions of the linear equation of nonstationary diffusion for an un-
     bounded body 17
  § 9. Solutions of the linear equation of nonstationary diffusion for a
     semibounded body... 19
  § 10. Solutions of the diffusion equation for a finite body 20
  § 11. Factors affecting the coefficient of diffusion 21
  § 12. Aspects of diffusion in rocks.. 23
  § 13. Methods of determining coefficients of diffusion in rocks......... 27
  § 14. Description of the diffusion of salt in rocks......................... 31
  § 15. Experimental data on the diffusion of salt in rocks 32
  § 16. The effect of moisture content on the diffusion of salt in rocks ... 35
  § 17. Experimental data on the diffusion of gases in rocks.............. 37
Literature cited.. 42

Chapter 3. Adsorption and Ion Exchange in the Interaction of Solutions and Gases
      with Rocks. 45
  § 18. The concept of adsorption 45
  § 19. Adsorption isotherms of gases and vapors on a homogeneous surface...... 46
  § 20. Adsorption of gases and vapors by porous adsorbents 47
  § 21. Adsorption of solutions on the surface of solids 48
  § 22. Ion exchange 49
  § 23. The concept of chemisorption 51
  § 24. Ion-exchange equilibrium in soils 52

§ 25. Investigation of adsorption and ion exchange in rocks   53
§ 26. Adsorption of gases on rocks   56
Literature cited   58

Chapter 4. The Kinetics of Adsorption, Ion Exchange, and Chemical Reactions
            of Solutions and Gases with Rocks   61
§ 27. The basic aspects of formal kinetics   61
§ 28. Heterogeneous processes   64
§ 29. Kinetic and diffusion regions of heterogeneous chemical reaction   65
§ 30. The kinetics of adsorption and ion exchange   66
§ 31. Equations of internal-diffusion kinetics of adsorption and ion ex-
      change   67
§ 32. The effect of temperature on reaction rate   70
§ 33. Methods of studying the kinetics of heterogeneous processes   72
§ 34. Experimental data on the kinetics of heterogeneous processes in
      rocks   73
Literature cited   74

Chapter 5. The Kinetics of Adsorption, Ion Exchange, and Chemical Reaction
            in a Current   75
§ 35. The kinetics of adsorption and ion exchange in the external-diffusion
      region   75
§ 36. The kinetics of adsorption and ion exchange in the internal-diffusion
      region   78
§ 37. The kinetics of adsorption and ion exchange with simultaneous con-
      sideration of intergranular and intragranular diffusion   80
§ 38. The kinetics of adsorption and ion exchange due to flow   81
§ 39. Determination of the diffusion mechanism controlling the rate of
      adsorption and ion exchange   82
§ 40. The kinetics of heterogeneous chemical reactions in a current   83
§ 41. Study of the kinetics of ion exchange in rocks from a current   86
Literature cited   90

Chapter 6. Geochemical Migration Due to Filtration and Diffusion   91
§ 42. Diffusion in a heterogeneous medium with consideration of adsorp-
      tion and ion exchange   91
§ 43. Diffusion in a heterogeneous medium, considering chemical reactions   97
§ 44. Filtration in a heterogeneous medium in the absence of interaction
      between substance and medium   98
§ 45. Filtration of a one-component solution in a porous medium, considering
      adsorption and ion exchanges   102
§ 46. Geochemical migration due to filtration in a weakly adsorbent medium   107
§ 47. Filtration in a heterogeneous medium with consideration of chemical
      reaction   111
§ 48. Filtration of a mixture of substances   113
§ 49. A dynamic method of determining the kinetic coefficients of adsorption
      and ion exchange   115
§ 50. Experimental results on filtration of solutions not interacting with the
      rocks   116
§ 51. A study of diffusion and filtration of adsorbable solutions and gases in
      rocks   117

§ 52. The so-called "filtration effect"   119
Literature cited.   123

Chapter 7. The Theory of Formation of Hydrothermal Deposits and Geochemical
            Aureoles at Deposits of Ores and Gas   125
      § 53. The formation of hydrothermal deposits and primary aureoles by
             interaction between solution and country rock   126
      § 54. The formation of deposits by reaction between components of hydro-
             thermal solutions..   136
      § 55. The theory of forming secondary geochemical aureoles of ore and gas
             deposits   140
      § 56. The question of oil formation   148
Literature cited   150

INTRODUCTION

The problem of the geochemical migration of elements has received wide attention in the works of V. I. Vernadskii and A. E. Fersman [1, 2]. Vernadskii considered geochemistry to be the science of the history of chemical elements on the earth, their distribution and movements in space and time, and their genetic relations [1]. Geochemical migration was defined by Fersman as "the movement of chemical elements in the earth's crust leading to their dissemination or concentration." The views of Vernadskii and Fersman on the migration of elements have received added support and development in connection with successes in physics, chemistry, biology, and other sciences.

According to Fersman, the earth is looked upon as a cosmic body, characterized by common origin and similarity of composition with the sun, the planets, meteorites, and other bodies of the solar system. The scale and trend of geochemical migration of elements in the earth are determined by the initial state of terrestrial matter, its thermal history, and the scale of time. The rules of elemental migration are determined by internal and external factors. Fersman distinguishes five groups of internal factors, i.e., factors related only to the properties of the atoms and their compounds: 1) the binding properties, including the physical constants of the substances; 2) the chemical properties, determining the reaction capability of the atoms and compounds; 3) the energy and crystallochemical properties of the substances; 4) the gravitational properties associated with the mass of the atoms; and 5) the radioactive properties of the atoms.

The external factors are those determined by the surrounding medium and do not depend on the individual properties of the migrating substances. According to Fersman, there are nine factors in the external group: 1) factors of cosmic migration, including gravitational and radiant energy, heat, pressure, electrical field, and the like; 2) factors of migration in melts, including conditions of gravitational equilibrium and gravitative differentiation; 3) factors of migration in aqueous solutions, including conditions of migration in solutions at high and low temperatures; 4) factors of migration in gas mixtures and supercritical solutions; 5) factors of mechanical migration; 6) factors of migration in colloidal and monocrystalline media; 7) factors of migration in the solid state; 8) factors of biogeochemical and industrial migration; and 9) other physicochemical factors.

The indicated classification of migration factors qualitatively embraces the principal types of elemental migration in the earth and is the theoretical basis for subsequent geochemical investigations. The logical development of the ideas of the founders of geochemistry — V. I. Vernadskii, V. M. Goldschmidt, and A. E. Fersman — must be transition from qualitative concepts and statistical interpretations to quantitative functional analysis of the geochemical processes of migration. Such transition, being characterized primarily by introduction of the time coordinate as an independent variable, is now possible through the theoretical and experimental achievements in scientific fields bordering geochemistry, above all in the field of physical chemistry. However, in geochemistry we have not yet attained sufficient development of the ideas of thermodynamically irreversible processes or of kinetically and dynamically physico-chemical processes having direct relation to the problem of geochemical migration. At present, experimental work is being carried out on the study of filtration and diffusion of solutions and gases, adsorption, and ion exchange in rocks. As a rule, this work is not connected with the problem of geochemical migration, but is conducted with other scientific and technical objectives. At the same time, the geochemical method of prospecting for mineral deposits is spreading constantly.

There is a certain lack of correspondence between practical requirements and the geologic sciences, on the one hand, and the theories and level of experimental work in the field of migratory processes, on the other. This disparity may be eliminated not only by means of experimental studies but also by constructing a single theory, embracing the principal migratory processes — filtration and diffusion — and by taking into account the fundamental processes of interaction between the migrating substances and the surrounding rocks: adsorption, ion exchange, and chemical reaction. The greatest significance for geochemistry is found in the heterogeneous processes of geochemical migration that take place at the interface between phases or that accompany the formation of new phases. The theoretical consideration of these heterogeneous processes of geochemical migration represents the substance of the present book.

The category of heterogeneous processes of geochemical migration is broader in content than the category of metasomatic processes. These latter are but a part of the heterogeneous processes of geochemical migration.

Let us pause briefly on the history of investigations in the field of geochemical migration. The first experiments on studying diffusion of salts in soils [3, 4, 5] were probably not even noticed by geologists. Several later views on diffusion, introduced by the geologist R. E. Liesegang [6], were immediately popularized in our country by Fersman [7].

In 1930 V. A. Sokolov [8] advanced the view of oil and gas surveys in prospecting for oil and gas deposits. On the basis of the fact that hydrocarbon gases migrate from deposits to the surface by means of diffusion and filtration, Sokolov suggested that one might determine the presence or absence of a deposit at depth by the presence or absence of hydrocarbon gases in subsoil air. This idea of Sokolov was not fully realized, chiefly because hydrocarbons are not formed only in connection with oil or gas deposits. Various concentrations of hydrocarbons are frequently encountered in sedimentary rocks at different depths, with or without oil or gas deposits being present, all yielding background concentrations in the subsoil air commensurate with the sought-for effect. Despite its inadequacy, the idea of oil and gas surveys played a positive role in geochemistry, since attempts to apply them on undisturbed structures stimulated the study of diffusion currents of gases in sedimentary rocks. Most important theoretical and experimental results in studying diffusion of hydrocarbon gases in sedimentary rocks have been obtained by P. L. Antonov [9-12]. However, he investigated only the simplest case, since the diffusion currents he studied were chemically indifferent to the rocks.

Among foreign geological publications, attention is called to the papers of Perrin and Roubault [13-17], in which the authors attempted a qualitative evaluation of the combination of molecular and, chiefly, ionic diffusion with ion-exchange reactions. They sought to explain the origin of metamorphic and igneous rocks by these processes. Perrin and Roubault could not use a quantitative approach in describing geologic processes because of the absence of any kinetic theory of heterogeneous reactions in a current, and also of any dynamic theory of solutions in porous media; these came into existence after the indicated work.

In our domestic geological literature, views on diffusion and filtration mass exchange of reagents in rocks were first discussed by Korzhinskii [18-21], who made substantial contributions to the study of metasomatic processes. He made a qualitative formulation of the dependence of the rate of metasomatism on the mechanism and rate of supply and removal of reagents in the rock. In evaluating the problem of metasomatism, Korzhinskii introduced into geology the view of systems with mobile and inert components and the view concerning the kinetic and thermodynamic method of studying metasomatic processes. Noting that diffusion and filtration are related to the kinetic method, Korzhinskii himself used the thermodynamic method chiefly in his studies of metasomatism. In explaining the mineral composition of rocks within the framework of the Gibbs' phase rule and the mineralogical phase rule of Goldschmidt, the thermodynamic method has a fundamental defect: by means of this method it is impossible to describe the occurrence in space and time of the processes of geochemical migration and mineral formation.

Korzhinskii formulated the empiric principle of differential mobility of elements in the earth's crust, a fundamental principle in geochemistry [19]. The quantitative foundation of this principle may be obtained in many cases on the basis of the theory of the kinetics and dynamics of chemical reactions (see Chaps. 6 and 7). The so-called "filtration effect," developed by Korzhinskii [20], is a qualitative theory of geochemical migration controlled by filtration, since it is based on the empirical principle of differential mobility of the chemical elements.

V. A. Zharikov [22-23] is a follower of Korzhinskii in the field of the thermodynamic method of investigating metasomatic processes. He attempts to describe metasomatic processes by using thermodynamically irreversible processes. In this, however, he uses the concept of the so-called "filtration effect," which does not take place in the general case (see [24] and, also, Chap. 6). The theory developed by Zharikov is therefore a qualitative theory of metasomatism.

The problem concerning the formation of geochemical aureoles, being a particular case of the heterogeneous processes of geochemical migration, was examined by R. I. Dubov [25]. Dubov, however, like the other authors, does not use equations of chemical kinetics in describing the process of forming aureoles.

The construction of a quantitative theory of heterogeneous processes of geochemical migration as made in this book is possible because of the extensive achievements in physical chemistry in the last 10-20 years, especially in the branches of kinetics and dynamics of adsorption, ion exchange, and chemical reactions. Fundamental contributions to the development of knowledge concerning the kinetics and dynamics of adsorption and ion exchange have been made by Martin, Boyd, Barrer, Glueckauf, Shilov, Dubinin, Zhukhovitskii, and others. The dynamical problem of chemical reactions was formulated by Panchenkov; for heterogeneous reactions it was solved by Thomas, Walter, and others.

The present book may be considered a first attempt at the step-by-step use of rates of chemical reactions in the science of geology [26]. It contains no exhaustive discussion of the questions of geochemical migration, but is an introduction to the theory of heterogeneous processes of geochemical migration. The theory developed has fundamental significance in such fields of science as mineralogy, hydrogeology, the science of oil and ore deposits, and may become the theoretical basis for geochemical methods of prospecting for mineral deposits.

LITERATURE CITED

1. Vernadskii, V. I., Collected Works [in Russian], Vol. 1, Izd. AN SSSR (1957).
2. Fersman, A. E., Geochemistry [in Russian], Vol. 2, Izd. AN SSSR, Moscow (1934).

3. Wollny, M. E., Vierteljahrsschr. der Bayrischen Landwirtschaft, Ergänzungsband, Heft I (1898).

4. Müntz, A., and Gaudechon, H., Annales de la Science Agronomique, I. f. 5Rt6, pp. 379–411 (1909).

5. Malpo; L., and Lefort, G., Annales de la Science Agronomique, p. 241 (1912).

6. Liesegang, R. E., Geologische Diffusionen, Dresden (1913).

7. Fersman, A. E., Priroda, No. 7-8, pp. 817-826 (1913).

8. Sokolov, V. A., The Gas Survey [in Russian], Moscow-Leningrad (1936).

9. Antonov, P. L., Neftyanoe Khozyaistvo, No. 5, p. 20 (1934).

10. Antonov, P. L., Geochemical Methods of Prospecting for Oil and Gas [in Russian], No. 15, Moscow (1953).

11. Antonov, P. L., Geochemical Methods of Prospecting for Oil and Gas [in Russian], No. 39, Moscow (1954).

12. Antonov, P. L., Geochemical Methods of Prospecting for Oil and Gas [in Russian], Moscow (1957).

13. Perrin, R., and Roubault, M., C. R. Acad. Sci., Vol. 227, No. 20, pp. 1043-1044 (1948).

14. Perrin, R., and Roubault, M., Soc. Geol. France, B. s. 5, Vol. 19, f. 1-3, pp. 3-14 (1949).

15. Perrin, R., C. R. Acad. Sci., Vol. 238, No. 17, pp. 1717-1720 (1954).

16. Perrin, R., C. R. Acad. Sci., Vol. 239, No. 21, pp. 1393-1395 (1954).

17. Perrin, R., C. R. Acad. Sci., Vol. 246, No. 21, pp. 2972-2976 (1958).

18. Korzhinskii, D. S., Izv. Akad. Nauk SSSR, Otdelenie Matematiki i Estestv. Nauk, No. 1, p. 35 (1936).

19. Korzhinskii, D. S., Zap. Vsesoyuzn. Mineralogich. Obshch., No. 71, p. 160 (1942).

20. Korzhinskii, D. S., Izv. Akad. Nauk SSSR, Ser. Geol., No. 2, p. 35 (1947).

21. Korzhinskii, D. S., Dokl. Akad. Nauk SSSR, Vol. 77, No. 2 (1951); Vol. 78, No. 1 (1951); Vol. 84, No. 4 (1952).

22. Zharikov, V. A., Geologiya Rudnykh Mestorozhd., No. 4, p. 3 (1965).

23. Zharikov, V. A., Geokhimiya, No. 10, p. 1191 (1965).

24. Garibyants, A. A., Golubev, V. S., and Beus, A. A., Izv. Akad. Nauk SSSR, Ser. Geol., No. 9, p. 26 (1966).

25. Dubov, R. I., in: The Geochemistry of Ore Deposits [in Russia], No. 117, Moscow (1964).

26. Panchenkov, G. M., and Lebedev, V. P., Chemical Kinetics and Catalysis [in Russian], Izd. MGU (1961).

CHAPTER 1

FORMULATING THE PROBLEM OF THE GEOCHEMICAL MIGRATION OF INCLUDED SUBSTANCES, AND METHODS OF SOLVING IT

§1. Formulating the Problem of Geochemical Migration of Included Substances

In general form, the problem of geochemical migration of included substances may be formulated in the following manner. Let there be a definite configuration of the environment (rock, soil) within which, or at the boundary of which, sources of migrating substances exist. We shall assume that at the moment tentatively adopted as zero (t = 0) the distribution of substances in the medium is known. As a consequence of migration, the distribution changes with time. The problem of the geochemical migration of included substances lies in determining the distribution function of the substances in the medium at any moment of time.

The problem cannot be solved in the general form because of mathematical difficulties. We must therefore seek some simplifications in formulating the problem. First, let us consider the migration of a single dissolved substance or of an individual gas. Second, let us assume that the host rock forms a homogeneous porous medium. This latter assumption means that any volume of the medium substantially greater than the size of the rock grains is characterized by constant averaged physical and chemical properties. Both these assumptions are found only as exceptions in nature. However, without a solution to the problem of migration of a single substance, it is impossible to consider the type of migration that is of most practical interest. The theory may be extended to more complex migration processes, as will be shown below. Setting up the present problem therefore has theoretical and practical interest.

A solution to the problem may be obtained by using the equation of material balance of the moving substance and the equation defining the interaction between the substance and the host rock with time. The latter equation is determined by the physicochemical laws of interaction between substances and the host rocks (sorption, ion exchange, chemical reaction) and is an equation of the kinetics of the corresponding physicochemical process.

§2. Equations of Material Balance and Kinetics of the Processes of Interaction between a Substance and Host Rocks

Let a liquid or gaseous solution, or an individual gas, move through the host rocks at a rate $\vec{u}$ (vector quantity). Let us assume that all the host rocks form a homogeneous porous medium; then the law of interaction between the dissolved substance and the medium is the same for any point in space. The composition of the solution will be characterized by the concentration C, the number of grams of dissolved substance in volume of solution per cm^3 of porous medium.

5

We introduce the rectangular coordinates (x, y, z) to represent the movement of the solution. The transfer of dissolved substance in the mobile phase is determined by two different mechanisms. First, when the concentration C is not the same throughout the bulk solution, diffusion of the dissolved substance occurs, causing thermal movement of the particles. As a result, the substance is transferred from the zone of high concentrations to a zone of low concentration (diffusion will be examined in more detail in Chap. 2). Second, the particles of dissolved substance are moved at rate $\vec{u}$ of the transporting current. The combination of these processes is normally called convective diffusion of a substance in a solution (gas) [1].

In the process of moving, the substance interacts with the rock: it is adsorbed by the rock and it enters into chemical reactions with the minerals of the rock. Consequently, the migrating substance is present also in an immobile phase, both as a compound identical to that in solution (adsorption) and as a new compound (chemical reaction). We shall use q to designate the number of grams of adsorbed substance or immobile phase through reaction with the rock in each cm^3 of porous medium. The substance in the solid phase does not ordinarily lose its mobility, but may diffuse through the volume of host rock [2].

As a result of these processes, the concentration of substance in the mobile and solid phases changes in space and time, so that C = C(x, y, z, t) and q = q(x, y, z, t) are functions of the coordinates and time. The problem of geochemical migration consists in finding these dependent relations.

Let us set up an equation for material balance for an arbitrary volume V of the porous medium. The imaginary surface S surrounding the volume V, in the normal case, cuts both free pore space and solid phase. Therefore the current $\vec{J}$ (vector quantity) of the substance, numerically equal to the amount of substance passing through 1 cm^2 of the surface S per second, is made up of the current of transport by convective diffusion in the mobile phase and the diffusion current in the immobile phase. The diffusion current $\vec{j_{\overline{D}}}$ in the immobile phase, in keeping with the law of Fick (Chap. 2), is equal to

$$\vec{j_{\overline{D}}} = -\overline{D}\,\mathrm{grad}\,q, \tag{1.1}$$

where $\overline{D}$ is the diffusion coefficient of the substance in the rock.

Since the amount of substance transported by a moving liquid or gas through 1 cm^2 of surface per second is equal to $\vec{Cu}$, the current $\vec{j_k}$ determined by convective diffusion is equal to

$$\vec{j_k} = \vec{Cu} - D\,\mathrm{grad}\,C, \tag{1.2}$$

where D is the diffusion coefficient of the substance in the bulk solution or gas.

Consequently

$$\vec{j} = \vec{j_{\overline{D}}} + \vec{j_k} = -D\,\mathrm{grad}\,C - \overline{D}\,\mathrm{grad}\,q + \vec{Cu}. \tag{1.3}$$

The amount of substance passing through the surface S per second is

$$Q = -\oint \vec{j}\,d\vec{S}, \tag{1.4}$$

where the integral is taken over the surface S surrounding the volume V. The direction from the surface outward was chosen as the positive direction of the vector of the exterior normal. The values $\partial C/\partial t$ and $\partial q/\partial t$ are changes in the amount of substance per second per unit volume of the immobile phases, respectively. Then, the change in amount of substance per volume V per second is

$$\int\limits_{(v)} \left(\frac{\partial C}{\partial t} + \frac{\partial q}{\partial t} \right) dV ,$$

where the integral is taken for the entire volume V. By virtue of the law of conservation, the change in amount of substance per volume V per second is equal to the amount of substance passing through the surface S per second:

$$\int\limits_{(v)} \left(\frac{\partial C}{\partial t} + \frac{\partial q}{\partial t} \right) dV = - \oint \vec{j}\, dS. \tag{1.5}$$

By transforming Eq. (1.5) by the Ostrogradskii–Gauss formula [3], we obtain

$$\int\limits_{(v)} \left(\frac{\partial C}{\partial t} + \frac{\partial q}{\partial t} \right) dV = - \int\limits_{(v)} \operatorname{div} \vec{j}\, dV. \tag{1.6}$$

Since the volume V was selected arbitrarily, Eq. (1.6) is valid only when the expressions under the integrals are equal. Consequently,

$$\frac{\partial C}{\partial t} + \frac{\partial q}{\partial t} = \operatorname{div}(D \operatorname{grad} C) + \operatorname{div}(\bar{D} \operatorname{grad} q) - \operatorname{div} C\vec{u}. \tag{1.7}$$

The diffusion coefficients D and $\bar{D}$ may be considered independent of the concentration of the solution (see Chap. 2). Then

$$\operatorname{div}(D \operatorname{grad} C) = D \operatorname{div} \operatorname{grad} C = D \Delta C,$$

where Δ is the Laplacian operator, in Cartesian coordinates:

$$\Delta = \frac{\partial^2}{\partial x^2} + \frac{\partial^2}{\partial y^2} + \frac{\partial^2}{\partial z^2} .$$

Expression (1.7) may be written in the form

$$\frac{\partial C}{\partial t} + \frac{\partial q}{\partial t} + \operatorname{div}(C\vec{u}) = D \Delta C + \bar{D} \Delta q. \tag{1.8}$$

Formula (1.8) is the equation of material balance of the dissolved substance in the process of its moving through the host rocks. In deriving Eq. (1.8) we did not consider the thermo-diffusion transfer of the substance, which is due to the presence of a temperature gradient in the medium, and other factors of migration. Thermal diffusion may be considered theoretically by introducing into the right side of Eq. (1.8) a supplementary term $\lambda \Delta T$ (where λ is the coefficient of thermal diffusion and T is the absolute temperature of the medium). However, because of the absence of experimental data on the transfer of substances by thermal diffusion in rocks, we will not consider this process.

The next equation characterizing geochemical migration is the kinetic equation of physicochemical interaction between the dissolved substance and the host rock, which defines the time of interaction of the substance with the medium at a fixed point (i.e., it gives an expression for the value of $\partial q / \partial t$). The principal processes of interaction between the solution and the host rock are sorption, ion exchange, and chemical reaction. The kinetic equations for these processes are examined in Chaps. 4 and 5. We shall write the kinetic equation in a general form without giving details of the interaction between the solution and the rock. In the most general case, the kinetics is determined by the concentration of the substance in the mobile and immobile phases, the rate constants [4] K_i of the chemical reactions between the solution and the medium (i = 1, 2, . . . is the number of reactions), the diffusion constants, and the flow rate. Mathematically, this dependence may be written in the implicit form

$$\frac{\partial q}{\partial t} = \varphi (C, q, K_i, \vec{u}, D, \bar{D}).$$

(1.9)

The explicit form of the function φ will be found below for each of the investigated processes.

The system of equations (1.8) and (1.9) characterizes the movement of the solution through the host rock. The flow rate $\vec{u}$ in the general case is a function of the coordinates and time. From the mathematical point of view, therefore, Eqs. (1.8) and (1.9) represent a system of differential equations with second-order partial derivatives and with variable coefficients. As follows from the theory of differential equations with partial derivatives [5], in order to find the concentration distributions C(x, y, z, t) and q(x, y, z, t), it is necessary to assign, in addition to Eqs. (1.8) and (1.9), a system of initial and boundary conditions for the investigated problem. The initial conditions give the concentration distributions q and C at the initial moment of time (t = 0). The boundary conditions define these functions at the boundaries of the investigated system and at the boundaries between phases. Depending on the nature of the physicochemical processes in the system, the initial and boundary conditions may be extremely diverse. They will be formulated separately for each actual problem.

The problem of geochemical migration for a single substance for the case when the host rocks do not form a homogeneous medium is formulated in analogous fashion. The system of equations (1.8) and (1.9) remains in force, but the rate constants K_i of the chemical reactions between solution and rock and the diffusion coefficients $\bar{D}$ and D will depend on the space coordinates.

§3. Hydrodynamic Equations

In Eqs. (1.8) and (1.9), which characterize the geochemical migration of included substances, the value of the flow rate $\vec{u}$ of the mobile phase is included. The rate $\vec{u}$ is commonly unknown, and only the forces acting on the liquid or gas in the process of moving are given. The rate $\vec{u}$ in the general case depends on the space coordinates and time: $\vec{u} = \vec{u}(x, y, z, t)$. Discovery of this dependence, when the active forces are known, is one of the hydrodynamical problems [1, 7].

Let us point out how we find a hydrodynamic solution to the problem of finding the function $\vec{u}(x, y, z, t)$.

The state of a moving liquid or gas is completely defined when, at each point in space, we know for any instant of time the three velocity components, u_x, u_y, and u_z, the density $\rho(x, y, z, t)$ of the fluid, and the pressure P(x, y, z, t). It is necessary to have a system of five equations in order to find these values.

The first equation (equation of continuity) follows from the law of conservation of matter and it has the form [2, 6]

$$\frac{\partial \rho}{\partial t} + \text{div}(\rho \vec{u}) = 0.$$

(1.10)

The following three hydrodynamic equations follow from Newton's second law of motion for an infinitesimally small volume of moving substance, bearing the name of the Navier–Stokes equation. In vector form the Navier–Stokes equation appears thus [1, 6]:

$$\frac{\partial \vec{u}}{\partial t} + (\vec{u}\,\mathrm{grad})\,\vec{u} = -\frac{1}{\rho}\,\mathrm{grad}\,P + \frac{\mu}{\rho}\,\Delta \vec{u} + \frac{\mu}{3\rho}\,\mathrm{grad}\,\mathrm{div}\,\vec{u} + \frac{\vec{f}}{\rho}, \tag{1.11}$$

where μ is the viscosity of the liquid and f is the body force acting on an element of liquid or gas.

On the left side of Eq. (1.11) stands the acceleration $\partial \vec{u}/\partial t$ of unit mass, and on the right are the acting forces. The first term on the right side accounts for the pressure gradient in the system; the second and third terms represent the effect of viscous forces [2] in the liquid or gas; the fourth term accounts for the external forces applied to the moving substance. The coordinate form of Eq. (1.11), particularly for the velocity projection on the x axis, has the form

$$\frac{\partial u_x}{\partial t} + u_x\frac{\partial u_x}{\partial x} + u_y\frac{\partial u_x}{\partial y} + u_z\frac{\partial u_x}{\partial z} = -\frac{1}{\rho}\frac{\partial P}{\partial x} + \frac{\mu}{\rho}\Delta u_x + \frac{\mu}{3\rho}\frac{\partial\,\mathrm{div}\,\vec{u}}{\partial x} + \frac{f_x}{\rho}, \tag{1.12}$$

where f_x is the projection of force f on the x axis.

Similarly, the equations for projections of u_y and u_z component velocities are recorded on the y and z axes.

The fifth equation is the equation of state of the liquid or gas, establishing the dependence of the density of the mobile phase on the pressure P and the concentration of dissolved substance C. This equation may be written in the general form

$$\rho = f(P, C). \tag{1.13}$$

Solution of the system of equations (1.10), (1.11), and (1.13) for definite initial and boundary conditions should yield the functional relations of flow rate, density of the mobile phase, and pressure on the space coordinates and time:

$$\vec{u} = \vec{u}(x, y, z, t); \quad \rho = \rho(x, y, z, t); \quad P = P(x, y, z, t).$$

The temperature T of the medium, which in the general case is a function of the coordinates and time, $T = T(x, y, z, t)$, does not enter Eqs. (1.10), (1.11), or (1.13), as it does not enter (1.8). Consequently, we neglect the dependence of liquid or gas density on temperature, and also the temperature dependence of the viscosity coefficient $\mu = \mu(T)$. As a first approximation, this may be done. Consideration of the temperature in the hydrodynamic equations complicates the problem greatly.

§4. Simplification and Methods of Solving the Problem

The system of equations of material balance for the dissolved substance (1.8), the kinetics of interaction between the substance and the medium (1.9), and the hydrodynamic equations (1.10), (1.11), and (1.13) characterize the geochemical migration of the dissolved substance, under definite initial and boundary conditions, without consideration of changes in the thermodynamic conditions of migration (pressure, temperature).

The system of equations is so complicated that it has not been possible to obtain a solution in analytical form. It has therefore become necessary to simplify the equations by introducing certain assumptions or to examine particular cases of the formulated problem. Let us examine the assumptions, from the physical point of view [7], that substantially simplify the problem of geochemical migration.

1. We shall assume that the mobile phase is incompressible and that its density is independent of the concentration of the dissolved substance (ρ = const). Then the equation of continuity (1.10) assumes the form

$$\operatorname{div} \vec{u} = 0. \tag{1.14}$$

Taking the equality of (1.14) into account, Eqs. (1.8) and (1.11) take on the forms

$$\frac{\partial C}{\partial t} + \frac{\partial q}{\partial t} + (\vec{u}\operatorname{grad})\,C = D\,\Delta C + \bar{D}\,\Delta q, \tag{1.15}$$

$$\frac{\partial \vec{u}}{\partial t} + (\vec{u}\operatorname{grad})\,\vec{u} = -\frac{1}{\rho}\operatorname{grad} P + \frac{\mu}{\rho}\,\Delta\vec{u} + \frac{\vec{f}}{\rho}. \tag{1.16}$$

The assumptions we have made are valid when the concentration of the dissolved substance is small (which apparently corresponds to the natural conditions of geochemical migration) and when change in density with change in concentration (as a result of adsorption, chemical reaction, etc.) may be neglected.

The system of equations (1.9) and (1.14)–(1.16), describing the migration of included substance with consideration of the assumptions we have made, is also complicated, and its solution in the general form has not been obtained.

2. In a porous medium there occurs a very complex distribution of velocities because of the inhomogeneity of the pores. It is difficult to account for the distribution of velocities in the pores by solving Eq. (1.14). Therefore, we normally introduce some average velocity $\vec{u} = \vec{u}(x,y,z,t)$ of movement of the solution at point (x,y,z) at time t.

Then, geochemical migration is described by the system of equations (1.9) and (1.15) at definite initial and boundary conditions. For the value of $\vec{u}$ we here take the average flow rate at the point with coordinates (x,y,z) at time t.

3. The diffusion coefficient $\bar{D}$ of elements and compounds in the solid phase (rock, soil) is several orders less than the diffusion coefficient $\bar{D}$ in the solution phase. Therefore, the diffusion current $j_{\bar{D}}$ in Eq. (1.3) may be neglected in comparison with $\vec{j}_k$. In this case, the equation of material balance takes on the form

$$\frac{\partial C}{\partial t} + \frac{\partial q}{\partial t} + (\vec{u}\operatorname{grad})\,C = D\,\Delta C. \tag{1.17}$$

Equations (1.9) and (1.17) represent the system we shall use below for describing the particular cases of migration of a single-component solution and gas in rocks and soils.

Methods of solving this system of equations are based on the theory of differential equations with partial derivatives [5]. The system of equations (1.9) and (1.17) is complex primarily because of the complexity of the kinetic equation (1.9). It therefore proves possible to find simple analytical solutions of the problem (by means of characteristics [5], operators [8], or other methods) only by considering individual cases of kinetics (kinetic equations of the first order). In other cases (kinetic equations of the second order; the interaction rate between solution and medium is determined by diffusion), it is possible to obtain only relatively simple asymptotic solutions (for time values of $t \rightarrow \infty$). In this regard there must be great value in computing the geochemical migration of substances by means of an electronic computer.

In the following chapters we shall consider solutions to the system of equations (1.9) and (1.17) for the following particular cases: 1) diffusion without consideration of the interaction between substance and rock (Chap. 6); 3) filtration without interaction between substance and rock (Chap. 6); and 4) filtration with consideration of interaction between substance and rock (Chaps. 5, 6, and 7).

§5. The Geochemical Migration of Mixtures

The problem of the geochemical migration of an n-component solution (n = 2, 3, 4 . . .) is formulated in analogous manner. The equations of material balance and the kinetics of the interaction between solution and host rock are written for each component of the mixture:

$$\frac{\partial C_i}{\partial t} + \frac{\partial q_i}{\partial t} + (\vec{u}\,\mathrm{grad})\,C_i = D_i\,\Delta C_i, \tag{1.18}$$

$$\frac{\partial q_i}{\partial t} = \varphi\,(C_1, C_2, \ldots, C_n;\, q_1, q_2, \ldots, q_n;\, K_i, \vec{u}, D_i), \tag{1.19}$$

where i = 1, 2, . . . , n represents the number of components of the system. The index i at all values in Eqs. (1.18) and (1.19) indicates that the values are referred to the i component.

The system of 2n equations (1.18) and (1.19) is complex primarily because of the complexity of the kinetic equation (1.19), since the rate of interaction of the i component with the rock, in the general case, depends on the concentrations of all n components. Solution of the problem of geochemical migration of mixtures has not been obtained in the general case.

The problem is substantially simplified when the interaction of each component of the solution with the host rock takes place independently of the other interactions. This condition is fulfilled when the solutions are not very concentrated. Then, the system of 2n equations (1.18) and (1.19) breaks down into n systems, each of which consists of two equations of the type

$$\frac{\partial C_i}{\partial t} + \frac{\partial q_i}{\partial t} + (\vec{u}\,\mathrm{grad})\,C_i = D_i\,\Delta C_i, \tag{1.20}$$

where

$$\frac{\partial q_i}{\partial t} = \varphi_i\,(C_i, q_i, K, \vec{u}, D_i), \text{ where } i = 1, 2, \ldots, n. \tag{1.21}$$

In the kinetic equation (1.21), in contrast to equation (1.19), concentrations of all components of the mixture, except that of the investigated component, are absent. The system (1.20) and (1.21) agrees with (1.9) and (1.17) (if the signs of i are omitted). This indicates that, for describing the migration of mixtures, we may use solutions to the problem of migration of single-component solutions, if the rate constants K and the diffusion coefficient D are replaced by the rate constants K_i and the diffusion coefficient D_i of the individual substances.

Below we shall examine concrete cases of geochemical migration of mixtures the descriptions of which may be made by transferring from the system (1.18) and (1.19) to (1.20) and (1.21). We point out first that this transition is more precise the more dilute the solution. For concentrations of solution of about 1N or greater, the system (1.20) and (1.21) only approximately defines the geochemical migration of mixtures.

LITERATURE CITED

1. Levich, V. G., Physicochemical Hydrodynamics [in Russian], Izd. Fiz.-Mat. Lit., Moscow (1959).
2. Seith, W., Diffusion in Metals [Russian translation], Izd. Inostr. Lit., Moscow (1958).
3. Bronshtein, I. N., and Semendyaev, K. A., Handbook of Mathematics [in Russian], Izd. Fiz.-Mat. Lit., Moscow (1959).

4. Panchenkov, G. M., and Lebedev, V. P., Chemical Kinetics and Catalysis [in Russian], Izd. MGU (1962).

5. Tikhonov, A. N., and Samarskii, A. A., Equations of Mathematical Physics [in Russian], Izd. Nauka, Moscow (1966).

6. Loitsyanskii, L. G., Mechanics of Liquids and Gases [in Russian], Gostekhizdat (1950).

7. Rachinskii, V. V., Introduction to the General Theory of Dynamics of Sorption and Chromatography [in Russian], Izd. Nauka, Moscow (1964).

8. Van der Pol, B., and Bremmer, H., Operational Calculus based on the Two-Sided Laplace Integral, Cambridge Univ. Press (1955).

CHAPTER 2

DIFFUSION IN ROCKS

The diffusion coefficients of the dissolved substances enter into Eqs. (1.9) and (1.17). Consequently, diffusion plays a fundamental role in the heterogeneous processes of geochemical migration. In the following chapters we shall examine the effect of diffusion on the range of geochemical migration (Chaps. 2, 6, and 7) and on the rate of interaction between the dissolved substance and the rock (Chaps. 4 and 5).

§6. Laws of Diffusion

Diffusion is the process of transferring a substance from one part of a system to another because of thermal movements of particles, molecules, atoms, ions, and the like. Diffusion is universal; it takes place both in individual substances and in any mixture of substances regardless of the aggregate state. In the first case the process is called self–diffusion, in the second, interdiffusion, or simply diffusion. The thermal movements of atoms and molecules are chaotic. Therefore, in an individual substance, diffusion confusedly transports particles from one site to another. However, when a system consists of two or more substances and the concentrations are not the same at different points, directional diffusion currents arise, tending to equalize the concentrations. This system by means of diffusion changes to a state of thermodynamic equilibrium, corresponding to the maximum disordered distribution of particles, i.e., to equality of concentration of each component in any part of the system. Consequently, diffusion is a spontaneous and irreversible process.

The laws of diffusion were first formulated by Fick in the form of two laws. The first of these states that the diffusion current of a substance is proportional to its concentration gradient and is directed toward the lessening of this gradient, or

$$\vec{j}_D = -D \operatorname{grad} C, \tag{2.1}$$

where $\vec{j}_D$ is the diffusion current, i.e., the amount of substance transferred by diffusion through a 1 cm^2 section of the medium per second, and D is the proportionality factor between the current and grad C, bearing the name coefficient of diffusion.

As follows from (2.1), the diffusion coefficient is numerically equal to the diffusion current when the concentration gradient is unity. Its dimensions are cm^2/sec. In rectangular coordinates (x, y, z) Eq. (2.1) takes on the form

$$j_D = -\left(D_x \frac{\partial C}{\partial x} + D_y \frac{\partial C}{\partial y} + D_z \frac{\partial C}{\partial z} \right), \tag{2.2}$$

where D_x, D_y, and D_z are the diffusion coefficients along the axes x, y, and z.

If the medium is isotropic, then

$$D_x = D_y = D_z = D. \tag{2.3}$$

If j_D changes with time, an accumulation or diminution of the diffusing substance occurs in the medium. Let us find the law of concentration change with time at a fixed point in the medium for the simplest case of one-dimensional diffusion along the x axis. Let us examine the diffusion in a cylinder with a cross-sectional area of 1 cm^2, the axis of which is parallel to the x axis. The change with time dt in the amount of substance within an infinitesimally small volume of the cylinder, formed by two planes perpendicular to x and separated by the distance dx, is equal to the difference between entering and departing currents, or

$$\frac{\partial C\,(x,\,t)}{\partial t}\,dx = j_D\,(x,\,t) - j_D\,(x+dx,\,t) = D\,\frac{\partial C\,(x,\,t)}{\partial x} + D\,\frac{\partial C\,(x+dx,\,t)}{\partial x}.$$

Expanding $C(x + dx, t)$ into a power series of dx, with dx tending toward zero, we obtain

$$\frac{\partial C\,(x,\,t)}{\partial t} = D\,\frac{\partial^2 C\,(x,\,t)}{\partial x^2}. \tag{2.4}$$

This equation is readily extended in the case of three-dimensional diffusion:

$$\frac{\partial C}{\partial t} = D_x\,\frac{\partial^2 C}{\partial x^2} + D_y\,\frac{\partial^2 C}{\partial y^2} + D_z\,\frac{\partial^2 C}{\partial z^2}. \tag{2.5}$$

If the medium is isotropic, then in place of Eq. (2.5) we have

$$\frac{\partial C}{\partial t} = D\left(\frac{\partial^2 C}{\partial x^2} + \frac{\partial^2 C}{\partial y^2} + \frac{\partial^2 C}{\partial z^2}\right) = D\,\Delta C, \tag{2.6}$$

where

$$\Delta = \frac{\partial^2}{\partial x^2} + \frac{\partial^2}{\partial y^2} + \frac{\partial^2}{\partial z^2}. \tag{2.7}$$

Equations (2.4) and (2.6) are mathematical expressions of Fick's second law in Cartesian coordinates. In solving these equations for definite initial and boundary conditions, it is possible to find the distribution of diffusing substance in the medium at any instant of time. Equations (2.4)-(2.6) were derived on the assumption that the diffusion coefficients are independent of the concentrations of the diffusing substance. In the general case, this assumption is invalid. Fick's second law must then be written in the form

$$\frac{\partial C}{\partial t} = \frac{\partial}{\partial x}\left[D\,\frac{\partial C}{\partial x}\right] + \frac{\partial}{\partial y}\left[D\,\frac{\partial C}{\partial y}\right] + \frac{\partial}{\partial z}\left[D\,\frac{\partial C}{\partial z}\right]. \tag{2.8}$$

Equation (2.6) is a particular case of the system of equations (1.9) and (1.17), describing the geochemical migration of substances under conditions when u = 0 and q = 0 (the absence of filtration and interaction between substances and rocks).

§7. Solutions of Diffusion Equations for Steady-State Current

The equation for steady-state diffusion, as follows from expression (2.6), has the form

$$D\,\Delta C = 0. \tag{2.9}$$

Equation (2.9) is a differential equation with partial derivatives of the second order. Therefore, in order to obtain a solution, it is necessary to supply six boundary conditions. In the case of one-dimensional diffusion, the number of boundary conditions is reduced to two.

Let us find the distribution law of steady-state concentrations $C(x, y, z)$ for simple configurations of the medium in which diffusion is taking place.

1. Plate. In the model of a plate, we assume that the length and breadth are considerably greater than the thickness, so that the effect of the edges may be neglected. In this case, the process is defined by the equation of one-dimensional diffusion:

$$\frac{d^2C}{\partial x^2} = 0, \tag{2.10}$$

where x is the coordinate along the height of the plate. Let the boundary conditions have the form

$$\begin{aligned} C &= C_1, \quad x = 0; \\ C &= C_2, \quad x = l. \end{aligned} \tag{2.11}$$

The general solution of Eq. (2.10) has the form

$$C(x) = Ax + B, \tag{2.12}$$

where A and B are the constants of integration.

The constants are found by solving Eq. (2.12) under the conditions of (2.11). In doing this, we obtain

$$\frac{C - C_1}{C_2 - C_1} = \frac{x}{l}. \tag{2.13}$$

The current through 1 cm^2 area of plate is equal to

$$j_D = -D\left(\frac{\partial C}{\partial x}\right)_{x=l} = D\frac{C_1 - C_2}{l}. \tag{2.14}$$

The amount of substance diffused through 1 cm^2 in time t is

$$Q = D\frac{C_1 - C_2}{l}t. \tag{2.15}$$

2. Cylindrical Tube. We shall place the z axis along the axis of the cylinder and introduce the cylindrical coordinates r, θ, such that [1]

$$\begin{cases} x = r\cos\theta, \\ y = r\sin\theta. \end{cases} \tag{2.16}$$

Equation (2.9) in cylindrical coordinates has the form

$$\frac{\partial C}{\partial t} = \frac{D}{r}\left[\frac{\partial}{\partial r}\left(r\frac{\partial C}{\partial r}\right) + \frac{\partial}{\partial\theta}\left(\frac{1}{r}\frac{\partial C}{\partial\theta}\right) + \frac{\partial}{\partial z}\left(r\frac{\partial C}{\partial z}\right)\right] = 0. \tag{2.17}$$

Let us assume that the tube is long, and that ring effects can therefore be neglected. Then, taking into account the symmetry of the problem, in place of Eq. (2.17) we have

$$\frac{1}{r}\frac{\partial}{\partial r}\left(r\frac{\partial C}{\partial r}\right) = 0. \tag{2.18}$$

The solution of Eq. (2.18) with the boundary conditions

$$C = C_1, \quad r = r_1,$$
$$C = C_2, \quad r = r_2 \tag{2.19}$$

is found in a manner analogous to the expression (2.13) and has the form

$$C = \frac{C_1 \ln \frac{r}{r_2} + C_2 \ln \frac{r_1}{r}}{\ln r_1/r_2}. \tag{2.20}$$

Hence, the amount of substance diffusing through unit length of cylinder in time t is

$$Q = \frac{2\pi D (C_2 - C_1)}{\ln r_2/r_1} t. \tag{2.21}$$

3. Hollow Sphere. Let us place the origin of the coordinate system at the center of the sphere and introduce the spherical coordinates (r, θ, ψ) [1]:

$$x = r \sin \theta \cos \varphi;$$
$$y = r \sin \theta \sin \varphi; \tag{2.22}$$
$$z = r \cos \varphi.$$

Then Eq. (2.9) is rewritten in the following form:

$$\frac{\partial C}{\partial t} = \frac{D}{r^2} \left[\frac{\partial}{\partial r} \left(r^2 \frac{\partial C}{\partial r} \right) + \frac{1}{\sin \theta} \frac{\partial}{\partial \theta} \left(\sin \theta \frac{\partial C}{\partial \theta} \right) + \frac{1}{\sin^2 \theta} \frac{\partial^2 C}{\partial \varphi^2} \right] = 0. \tag{2.23}$$

Let us find the solution of the problem of steady-state diffusion for the boundary conditions

$$C = C_1, \quad r = r_1;$$
$$C = C_2, \quad r = r_2. \tag{2.24}$$

In view of the symmetry of the distribution of diffusing substance in the hollow sphere, the equation is written

$$\frac{1}{r} \frac{\partial^2 (rC)}{\partial r^2} = 0. \tag{2.25}$$

The solution of Eq. (2.25) under conditions (2.24) takes on the form

$$\frac{C_1 - C}{C_1 - C_2} = \frac{r_2 (r - r_1)}{r (r_2 - r_1)}. \tag{2.26}$$

The current through the entire surface of the sphere is

$$j_D = -4\pi r_2^2 D \left(\frac{\partial C}{\partial r} \right)_{r=r_2} = 4\pi D (C_1 - C_2) \frac{r_2 r_1}{r_2 - r_1}. \tag{2.27}$$

Similarly we may find solutions of other problems. Some of them will be considered in Chap. 7 in connection with the formation of geochemical dissemination aureoles. We may note that the solutions of many problems of steady-state diffusion may be borrowed from the well-developed theory of thermal conductivity [2–3], since Eq. (2.9) is similar to the equation of steady-state thermal conductivity.

§8. Solutions of the Linear Equation of Nonstationary Diffusion for an Unbounded Body

The solution of the linear equation of nonstationary diffusion

$$\frac{\partial C}{\partial t} = D \frac{\partial^2 C}{\partial x^2} \tag{2.28}$$

may be divided into solutions for infinite, semiinfinite, and finite bodies. Let us examine the general solution of Eq. (2.28) for an infinite isotropic body, extending from $-\infty$ to $+\infty$ in the direction of the x axis, with the initial concentration distribution in the body given in the form

$$C(x, 0) = f(x). \tag{2.29}$$

We shall seek a solution by separation of the variables [1]

$$C(x, t) = X(x) T(t), \tag{2.30}$$

where X and T are functions depending only on the variables x and t respectively.

By substituting Eq. (2.30) in equation (2.28) and integrating under conditions (2.29), it is not difficult to find the desired solution [4-6]:

$$C(x, t) = \frac{1}{2\sqrt{\pi D t}} \int_{-\infty}^{+\infty} f(\xi) e^{\frac{-(\xi - x)^2}{4Dt}} d\xi, \tag{2.31}$$

where ξ is the variable of integration.

Formula (2.31) is the general solution of Eq. (2.28) for an isotropic unbounded body. Let us find some particular solutions of the problem of nonstationary diffusion for different boundary conditions, using Eq. (2.31).

1. Diffusion from a Semiinfinite Space. Let the initial concentration distribution in the unbounded body (extending from x = $-\infty$ to x = $+\infty$) be given in the form

$$C(x, 0) = \begin{cases} C_0, & x < 0 \\ 0, & x > 0. \end{cases} \tag{2.32}$$

By splitting the integral in Eq. (2.31) into two (from $-\infty$ to zero and from zero to $+\infty$) and accounting for conditions (2.32), we obtain

$$C(x, t) = \frac{C_0}{2} \left[1 - \frac{2}{\sqrt{\pi}} \int_{0}^{x/2\sqrt{Dt}} e^{-\xi^2} d\xi \right] = \frac{C_0}{2} \left[1 - \mathrm{erf}\left(\frac{x}{2\sqrt{Dt}} \right) \right], \tag{2.33}$$

where

$$\mathrm{erf}\, z = \frac{2}{\sqrt{\pi}} \int_{0}^{z} e^{-y^2}\, dy \tag{2.34}$$

is the integral distribution of Gauss.

This function has been tabulated, so that the distribution (2.33) for any x and t may be easily computed. The distribution for different moments of time is shown in Fig. 1.

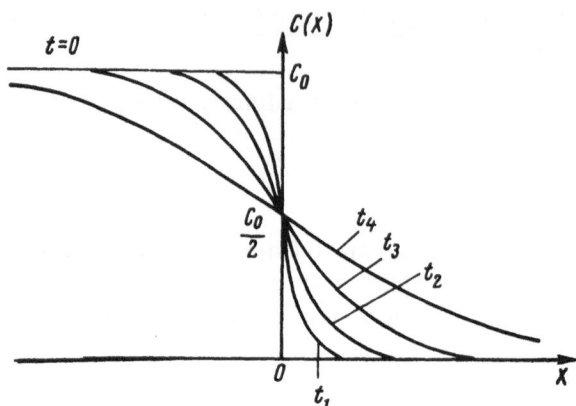

Fig. 1. Concentration distribution during diffusion from a semiinfinite space.

In order to determine the range of the diffused substance from the semiinfinite space, it is possible, by expanding the Gaussian integral into a power series and limiting ourselves to $C/C_0 \geq 0.15$ in the first member of the series, to write

$$C\,(x,\,t) \approx \frac{C_0}{2}\left[1 - \frac{x}{\sqrt{\pi Dt}}\right].\tag{2.35}$$

We shall use $C_{\min}$ to designate the minimum concentration of diffusing substance that may still be determined by quantitative analyses. The distance the diffused substance has penetrated will be indicated by $x_{\max}$, where the concentration of diffused substance is $C_{\min}$. From expression (2.35) it follows that

$$x_{\max} = 2\left(1 - \frac{2C_{\min}}{C_0}\right)\sqrt{\pi Dt}.\tag{2.36}$$

If $C_{\min} = \alpha C_0$, then

$$x_{\max} \approx \sqrt{\pi Dt}\ \ (1 - 2\alpha).\tag{2.37}$$

By using Eqs. (2.36) and (2.37), we may calculate the maximum distance the diffusing substance will reach after a given time.

2. Diffusion from an Infinitesimally Thin Layer. Let the initial concentration distribution in the unbounded body be given in the following form:

$$C\,(x,\,0) = \begin{cases} C_0, & x_0 - l < x < x_0 + l \\[4pt] 0, & \begin{aligned} &x < x_0 - l \\ &x > x_0 + l, \end{aligned} \end{cases}\tag{2.38}$$

where x_0 is the coordinate of a fixed point and l is a small value.

By integrating Eq. (2.31) between the limits ($x_0 - l$; $x_0 + l$), it is easy to show [4–6] that

$$C\,(x,\,t) = \frac{Q}{2\sqrt{\pi Dt}}\, e^{\frac{-(\lambda l - x)^2}{4Dt}}\ ,\tag{2.39}$$

where the parameter $-1 < \lambda < +1$; $Q = C_0\,2l$ is the initial amount of substance in the layer ($x_0 - l$, $x_0 + l$).

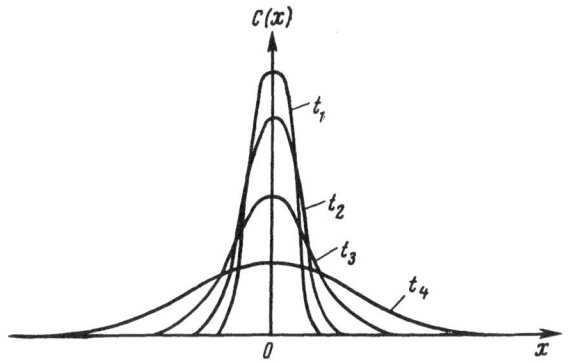

Fig. 2. Concentration distribution during diffusion from an infinitesimally thin layer in an unbounded body.

When l tends toward zero (i.e., when we consider diffusion from an infinitesimally thin layer), we obtain an instantaneous source of substance at the point x_0, which, at $t > 0$, gives a concentration distribution in the form

$$C(x, t) = \frac{Q}{2\sqrt{\pi Dt}} e^{-\frac{(x_0-x)^2}{4Dt}}. \tag{2.40}$$

The distribution for the case of the source being placed at the origin is shown in Fig. 2 for different moments of time. The distribution is symmetrical relative to the maximum concentration, the value of which, as follows from Eq. (2.40), is

$$C_{max} = \frac{Q}{2\sqrt{\pi Dt}}, \tag{2.41}$$

whence it may be seen that C_{max}, decreases with time proportionally to $\sqrt{t}$.

§9. Solutions of the Linear Equation of Nonstationary Diffusion for a Semibounded Body

By a semibounded body we mean one that is bounded on only one side (the plane $x = 0$), extending to infinity on the other. The general solution of Eq. (2.28) for the initial condition

$$C(x, 0) = f(x), \quad x > 0 \tag{2.42}$$

is found from Eq. (2.31). It has the form

$$C(x, t) = \frac{1}{2\sqrt{\pi Dt}} \int_0^\infty \left[f(\xi) e^{-\frac{(\xi-x)^2}{4Dt}} + f_1(-\xi) e^{-\frac{(\xi+x)^2}{4Dt}} \right] d\xi, \tag{2.43}$$

where $f_1(-\xi)$ is a function found from the boundary conditions.

In particular, when we examine diffusion from a steady source, i.e., under the conditions

$$C(0, t) = C_0,$$
$$C(x, 0) = 0, \tag{2.44}$$

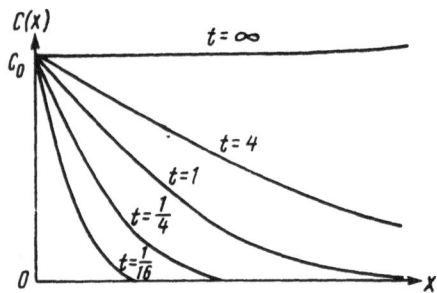

Fig. 3 Concentration distribution during diffusion from a steady source in a semibounded body.

we obtain from Eq. (2.43) [4-6]

$$C(x, t) = C_0 \left[1 - \text{erf} \left(\frac{x}{2\sqrt{Dt}} \right) \right]. \qquad (2.45)$$

The distribution of substance diffusing from the steady source is shown graphically in Fig. 3 for different moments of time.

§ 10. Solutions of the Diffusion Equation for a Finite Body

We shall present some solutions of the equation of nonstationary diffusion (2.6) for finite bodies, which we shall need in our further discussion.

1. **Plate with a Reflecting Face.** Let us place the origin at the lower surface of the plate and assume that at zero time no diffusing substance is present in the plate. The boundary x = 0 is reflecting; i.e., it is impermeable for the substance. The initial and boundary conditions corresponding to diffusion from a steady source have the form

$$t = 0, \quad 0 < x < l, \quad C = 0,$$
$$t > 0, \quad x = l, \quad C = C_0, \qquad (2.46)$$

$$t > 0, \quad x = 0, \quad \frac{\partial C}{\partial x} = 0 \quad \text{(absence of current through the boundary x = 0)}.$$

The solution of Eq. (2.28) with conditions (2.46) may be written in the form [2]

$$\frac{C}{C_0} = 1 - \frac{4}{\pi} \sum_{n=0}^{\infty} \frac{(-1)^{n+1}}{(2n+1)} \exp\left[-(2n+1)^2 \frac{\pi^2 Dt}{4l^2} \right] \cos(2n+1) \frac{n\pi x}{2l}, \qquad (2.47)$$

where n = 0, 1, 2, 3, . . .

Numerical calculations of the distribution of the substance in the plate may be made by investigating the convergence of the series and the summed members of the series (2.47).

The average concentration $\overline{C}(t)$ in the plate at any moment of time is computed from the formula

$$\overline{C} = \frac{1}{l} \int_0^l C(x, t)\, dx \qquad (2.48)$$

and is equal [2] to

$$\frac{\overline{C}}{C_0} = 1 - \frac{8}{\pi^2} \sum_{n=0}^{\infty} \frac{1}{(2n+1)^2} \exp\left\{ -\frac{(2n+1)^2 \pi^2 Dt}{l^2} \right\}. \qquad (2.49)$$

Values of $\overline{C}/C_0$ for different values of Dt/l^2 have also been tabulated [8].

Asymptotic investigation of Eq. (2.49) shows that, for values of $\overline{C}/C_0 < 0.5$, the distribution of diffusing substance in the medium may be shown approximately in the form

$$\frac{\overline{C}}{C_0} = \sqrt{\frac{\pi^2 D}{l^2}\, t}. \qquad (2.50)$$

Hence, the maximum thickness of the plate l_{max} through which the substance during diffusion may still be detected at time t by quantitative analysis is

$$l_{max} = \pi \frac{C_0}{C_{min}} \sqrt{\overline{Dt}}.$$ (2.51)

The value l_{max} describes the distance the diffusing substance will penetrate after time t during diffusion into a plate with a reflecting face.

2. Sphere. Let us place the origin at the center of the sphere and assume the sphere to be isotropic. Then the equation of nonstationary diffusion in spherical coordinates has the form

$$\frac{\partial C}{\partial t} = D \left(\frac{\partial^2 C}{\partial r^2} + \frac{2}{r} \frac{\partial C}{\partial r} \right).$$ (2.52)

Let us assume that the sphere, in which no diffusing substance is present at zero time, is placed in a medium of the substance (solution or gas) with uniform concentration C_0. The initial and boundary conditions of this problem have the form

$$C(r_0, t) = C_0,$$

$$C(r, 0) = 0,$$ (2.53)

$$\frac{\partial C(0, t)}{\partial r} = 0 \text{ (symmetry conditions)}.$$

The solution of Eq. (2.52) under conditions (2.53) may be written in the following form

$$\frac{C}{C_0} = 1 - \frac{2}{\pi} \sum_{n=1}^{\infty} \frac{(-1)^{n+1}}{n} \frac{r_0}{r} \sin \left(n\pi \frac{r}{r_0} \right) \exp \left(- \frac{n^2 \pi^2}{r_0^2} Dt \right).$$ (2.54)

The average concentration $\overline{C}$ in the sphere is found by the formula

$$\overline{C} = \frac{3}{r_0^3} \int_0^{r_0} r^2 C(r, t) \, dr$$ (2.55)

and is equal [9] to

$$\frac{\overline{C}}{C_0} = 1 - \frac{6}{\pi^2} \sum_{n=1}^{\infty} \frac{1}{n^2} \exp \left(- \frac{n^2 \pi^2}{r_0^2} Dt \right).$$ (2.56)

Tables of values of the function $\overline{C}/C_0 = f(Dt/r_0^2)$ are given in the monograph of Lykov [3].

In conclusion, we note that solutions of many problems of nonstationary diffusion may be borrowed from the theory of thermal conductivity [2-3], since Eq. (2.6) is analogous to the equation of nonstationary thermal conductivity.

§ 11. Factors Affecting the Coefficient of Diffusion

The coefficient of diffusion depends on temperature, pressure, concentration of the diffusing particles, and the mass of these particles. We shall examine these relations separately for gases, liquids, and solids.

Theoretical expressions for coefficients of diffusion and self diffusion of gases are found from the molecular-kinetic theory of gases [10-11]. These expressions are different according to the molecular model of gas employed (rigid elastic spheres, rough elastic spheres, repelling or attracting spheres). However, experimental data are inadequate for us to give preference to any of the gas models. From all the theories, it follows that the coefficient of diffusion depends on pressure and temperature in the following way [10]:

$$D \approx \frac{T^{1+S}}{P},\tag{2.57}$$

where $1/2 \le S \le 1$ and depends on the gas model.

The dependence (2.57) is realized in experiment, in which S lies between the extreme values of $1/2$ and 1. Thus, for interdiffusion in the following mixtures: $CO_2 - air$, S = 0.968; $H_2 - O_2$, S = 0.755; $CO_2 - H_2$, S = 0.742; $O_2 - N_2$, S = 0.792 [10]. The dependence of the diffusion coefficient of gases on concentration and molecular weight is complex [10-11]. Dependence of the diffusion coefficient on concentration is generally slight, and it can frequently be neglected. With increasing molecular weight M of the gas, the diffusion coefficient declines. As a first approximation, it may be stated that the diffusion coefficient is inversely proportional to $\sqrt{M}$.

There are different expressions for the diffusion coefficient in liquids. We shall cite the theoretical expression for the coefficient of self diffusion of liquids obtained by Panchenkov [12]:

$$D = \frac{4\sqrt[3]{3}\sqrt{2R}}{\sqrt[3]{4\pi N_0}\sqrt{\pi}} \frac{V_m^{1/3} T^{1/2}}{M^{1/2}} e^{-\frac{\lambda_i}{\gamma RT} + \frac{\varepsilon_0}{2RT}} \left[\frac{\varepsilon_0}{2RT} + 1 \right],\tag{2.58}$$

where ε_0 is the energy of a single molecular bond in the liquid, λ_i the molecular heat of vaporization at the given temperature T, γ the coordination number of the liquid, V_m the gram-molecular volume, N_0 Avogadro's number, and R the universal gas constant.

From expression (2.58) it is seen that the coefficient of self diffusion of a liquid is inversely proportional to the square root of the molecular weight. The dependence of D on temperature is complex. As a first approximation we may write, in place of Eq. (2.58) [12],

$$\ln D = \ln A + \frac{\alpha \ln T}{R} - \frac{\varepsilon_0}{RT},\tag{2.59}$$

where A and α are constants.

If the investigated temperature interval is small, then, by expanding log T into a series and restricting ourselves to two terms of the series, we find that the dependence of the diffusion coefficient on $1/T$ is exponential.

The theory of diffusion in the solid phase was first developed by Frenkel [13]. According to this theory, and also to later work [14], the temperature dependence of the diffusion coefficient of a substance in the solid phase has the form

$$\overline{D} = D_0 e^{-\frac{E}{RT}},\tag{2.60}$$

where E is the activation energy of diffusion, i.e., that energy a particle must have in order to move from one position of equilibrium in a crystal lattice to another. The physical significance of the factor D_0 in front of the exponent has been interpreted differently by a number of authors.

From the theory of absolute rates of reactions [14], it follows that the diffusion coefficient of an isotope in the solid phase is inversely proportional to the square root of its mass, and this is generally confirmed by experiment.

§12. Aspects of Diffusion in Rocks

Diffusion in rocks only in exceptional cases conforms to the rules indicated above, since it takes place in a more complex environment than that assumed in deriving Eq. (2.4) and all the succeeding ones. All rocks are porous; they contain pores of various sizes and shapes, and the mechanism of transfer of substance in them is very complex [15-27]. Rocks (sand, clay, and other deposits) are heterogeneous systems, containing solutions or gases, or solutions and gases at the same time. During diffusion in a heterogeneous medium, the substance interacts with the rocks (is adsorbed, exchanges ions, enters into chemical reactions). Under natural conditions, the substance normally diffuses through a series of unlike beds, in which the diffusion coefficient varies, and this complicates our description of the diffusion process.

Let us examine ways that we might take into account the basic factors indicated in describing diffusion in rocks.

1. Porosity of the Rocks. The transport of a substance in a porous body takes place through pores and is a complex phenomenon. The transport of a substance in the liquid phase is controlled by molecular diffusion. The processes of transport in the gas phase are more varied. If the diameter of the pores d is greater than λ, the length of the free path of the gas molecules, $(\lambda/d \ll 1)$, the transport is controlled by ordinary diffusion in the volume of the gas. At very low pressures in the gas or in tight pores $(\lambda/d \gg 1)$, the molecules for the most part do not collide with each other (as in the case $\lambda/d \ll 1$) but with the walls of the pores. The mechanism of transporting the substance here is different, and is called Knudsen diffusion. Lastly, in the ultrapores, the size of which is comparable with the cross section of the molecules, so-called "zeolitic diffusion" takes place [9], and is very sensitive to the size of the diffusing molecules.

Since rocks have pores of all sizes and shapes, the transport of the substance takes place by all the methods indicated above at the same time. This makes it diffucult to furnish a quantitative description of the process. Theoretical investigations have shown [9] that the rate of transport in all cases has the same dependence on concentration gradient, analogous to Fick's laws of diffusion (2.1) and (2.6). Consequently, with any type of transport of a substance through a porous medium, including transport through rocks, the rate of the process may be formally expressed by the diffusion equations (2.1) and (2.6) with some effective diffusion coefficient.

2. Structure of the Rock. Let us examine diffusion in granular rocks. Diffusion of dissolved substances in this case may take place in free space in the rock (i.e., in the interspaces between grains). If the spaces between grains are filled with gas, then only diffusion of gases is possible. When the free space of the rock is partly or completely filled with water, then diffusion of dissolved substances in the liquid phase takes place along with diffusion of gases.

Let us consider a layer of rock with a thickness l in which a linear current of diffusing substance has been established. Because of the heterogeneity of the medium, the true path of particles diffusing through the rock is $\eta' l$, where $\eta' > 1$ is a value that indicates how many times greater the path of particle movement is than l. The flow of diffusing substance through 1 cm^2 of substance under conditions that the concentration refers to unit volume of rock [cf. Eq. (2.14)] is

$$j_D = D \frac{c_1 - c_2}{l} = D_0 \frac{c_1 - c_2}{\eta' l}, \qquad (2.61)$$

where D is the diffusion coefficient in the rock, and D_0 the diffusion coefficient in the liquid or gas filling the interspaces between grains.

From Eq. (2.61) it follows that

$$D = D_0 \frac{1}{\eta'} = \eta D_0 \ (\eta < 1), \tag{2.62}$$

where η (frequently η') is the tortuosity factor.

Thus, the diffusion coefficient of the substance D in the free space of a heterogeneous medium, filled with solution or gas, is smaller than the corresponding diffusion coefficient D_0 in the bulk solution or gas. The tortuosity factor depends on the packing of the particles, but does not depend on the particle size (the geometric shape of the particles remaining unchanged). Thus, for cubic packing of spherical particles of identical sizes, $\eta' = \pi/2$, but for rhombic packing, $\eta' = 2\pi/3\sqrt{3}$ [15].

When the concentration of diffusing substance is referred to unit volume of solution, then, in place of expression (2.61), we have

$$j_D = D \frac{C_1 - C_2}{l} = D_0 \varkappa \frac{C_1 - C_2}{\eta' l}, \tag{2.63}$$

where $\varkappa$ is the porosity of the medium. The correlation between the diffusion coefficient D in the porous medium and D_0 in the solution or gas is then expressed by

$$D = \eta \varkappa D_0. \tag{2.64}$$

In Eqs. (1.9) and (1.17), describing the geochemical migration of dissolved substances, the diffusion coefficient D does not depend on the porosity of the medium, since in it the concentrations C and q refer to unit volume of the porous medium.

The substance diffusing through rocks is generally adsorbed or it exchanges ions and enters into chemical reaction with the substances of the rock (see Chap. 3). These processes are not taken into account in the derivation of the equation for nonstationary diffusion (2.4) and (2.8). Consequently, the solutions of these equations cannot describe diffusion of substances in rocks. This was noted long ago [16-19], and was interpreted as failure of diffusion in rocks to conform to Fick's law. Diffusion accompanied by interaction between substance and rock is defined, as pointed out in Chap. 1, by the system of the equation of material balance and the kinetic equation for the interaction between substance and medium (for more detail see Chap. 6).

3. Moisture Content of the Rock. The moisture content of rocks has been shown to have a substantial effect on the diffusion rate of dissolved substances and gases [16-22]. The diffusion of gas takes place through the air filling the pores. When wetting of the rock is incomplete, water fills some of the pores, chiefly the capillary pores. Consequently, with increase in moisture content the diffusion rate of gases declines. In damp and wet rocks, diffusion takes place chiefly through the noncapillary (interaggregate) pores. When gas is dissolved in the water, the gas diffuses in the liquid phase. However, when the solubility of the gases is low, it may be assumed that diffusion transfer of dissolved gas is negligibly small.

Quantitatively the effect of moisture content on diffusion rate of gas may be taken into account if we know the dependence of the diffusion coefficient on the relative moisture content of the rock $D = D(W)$ and the law of moisture distribution in the rock $W = W(x, y, z)$. By knowing these relations, we may find how the diffusion coefficient changes from point to point, $D \neq D(x, y, z)$. Diffusion may then be described by means of Fick's laws (2.1) and (2.8) with diffusion coefficients depending on the space coordinates $D = D(x, y, z)$. The problem becomes analogous to the problem of diffusion in inhomogeneous media.

It is a more complex matter to determine the role of moisture in the diffusion of substances dissolved in the liquid phase (electrolytes and nonelectrolytes). The diffusion of dissolved substances may take place only when the pores of the rock are partially or completely filled with water. The lower the specific content of water in the pores, the lower the diffusion rate. However, the phenomenon of diffusion is complicated by the osmotic transfer of water from a site with lower concentrations of the dissolved substance to sites of higher concentrations. Thus, osmotic transfer of water takes place in a direction opposite to that of diffusion, which leads to decrease in the concentration gradient and, consequently, to a decline in diffusion rate. The role of moisture in diffusion of dissolved substances will be discussed in more detail in Section 11 of the present chapter.

4. Stationary Diffusion in Multilayered Media. The problem of one-dimensional and two-dimensional stationary diffusion through a system of n parallel beds with thicknesses of h_1, h_2, ..., h_n, a summed thickness of H, and diffusion coefficients D_1, D_2, ..., D_n has been investigated by Antonov [23]. We set the x axis perpendicular to the interface of the beds and the y axis along the interface. The solution of the equation of one-dimensional diffusion

$$\frac{\partial}{\partial x}\left[D(x)\frac{\partial C}{\partial x}\right]=0, \tag{2.65}$$

where

$$
\begin{aligned}
D(x) &= D_1, & 0 &< x < h_1, \\
D(x) &= D_2, & h_1 &< x < h_1+h_2, \\
&\cdots\cdots\cdots\cdots\cdots \\
D(x) &= D_n, & h_{n-1} &< x < h_n,
\end{aligned}
\tag{2.66}
$$

with the boundary conditions

$$
\begin{cases}
x=0, & C=0, \\
x=H, & C=C_0
\end{cases}
\tag{2.67}
$$

may be written in the form

$$C(x)=C_0\frac{x}{H_{ef}}, \tag{2.68}$$

where

$$
\begin{cases}
H_{ef}=h_1\frac{D_1}{D_1}+h_2\frac{D_1}{D_2}+\ldots+h_n\frac{D_1}{D_n}, & 0<x<h_1, \\
H_{ef}=h_1\frac{D_2}{D_1}+h_2\frac{D_2}{D_2}+\ldots+h_n\frac{D_2}{D_n}, & h_1<x<h_1+h_2, \\
\cdots\cdots\cdots\cdots\cdots\cdots\cdots\cdots\cdots \\
\quad H_{ef}=h_1\frac{D_n}{D_1}+h_2\frac{D_n}{D_2}+\ldots+h_n\frac{D_n}{D_n}, \\
h_1+h_2+\ldots+h_{n-1}<x<h_1+h_2+\ldots h_n.
\end{cases}
\tag{2.69}
$$

Equation (2.68) agrees with expression (2.13) when we substitute the value l = H = $h_1+h_2+\ldots+h_n$ for H_{ef}. Thus, the concentration distribution of diffusing substances in the system of n parallel beds of height H agrees with the concentration distribution in a homogeneous medium with height H_{ef}.

The problem of nonstationary diffusion in an inhomogeneous medium is complex, and it has not been solved for the general case. We shall examine solutions of two particular problems, borrowed from the theory of thermal conductivity [2]:

5. Diffusion in an Unbounded Compound Body. Let the diffusion coefficient of a substance in the region $x > 0$ be D_1, and in the region $x < 0$ let it be D_2; and let there be no diffusing substance in the region $x < 0$ at zero time. The concentration distribution $C_1(x,t)$ in the region $x > 0$ and $C_2(x,t)$ in the region $x < 0$ is found by solution of the differential equations

$$\frac{\partial C_1(x,\,t)}{\partial t} = D_1 \frac{\partial^2 C_1(x,\,t)}{\partial x^2} \,, \tag{2.70}$$

$$\frac{\partial C_2(x,\,t)}{\partial t} = D_2 \frac{\partial^2 C_2(x,\,t)}{\partial x^2} \,, \tag{2.71}$$

with the initial conditions

$$t = 0, \quad C_1(x,\,t) = C_0, \quad C_2(x,\,t) = 0 \tag{2.72}$$

and the conditions for conjugate relations

$$x = 0,\ t > 0,\ C_1 = C_2, \quad \text{(equality of concentrations)}$$

$$D_1 \frac{\partial C_1}{\partial x} = D_2 \frac{\partial C_2}{\partial x} \quad \text{(equality of currents)}. \tag{2.73}$$

A solution of (2.43) is achieved in each of the regions, so that

$$C_1 = A_1 + B_1 \operatorname{erf}\left(\frac{x}{2\sqrt{D_1 t}}\right), \qquad x > 0, \tag{2.74}$$

$$C_2 = A_2 + B_2 \operatorname{erf}\left(\frac{|x|}{2\sqrt{D_2 t}}\right), \qquad x < 0. \tag{2.75}$$

The constants of integration A_1, B_1, A_2, and B_2 are obtained from Eqs. (2.74) and (2.75) under conditions (2.72) and (2.73). Finally we obtain

$$C_1 = C_0 \frac{D_1^{1/2}}{D_1^{1/2} + D_2^{1/2}} \left\{ 1 + \frac{D_2^{1/2}}{D_1^{1/2}} \operatorname{erf}\left(\frac{x}{2\sqrt{D_1 t}}\right) \right\}, \tag{2.76}$$

$$C_2 = C_0 \frac{D_1^{1/2}}{D_1^{1/2} + D_2^{1/2}} \left\{ 1 - \operatorname{erf}\left(\frac{|x|}{2\sqrt{D_2 t}}\right) \right\}. \tag{2.77}$$

6. Diffusion in a Semibounded Compound Body. Let us examine the region $-l < x < \infty$ in which at $-l < x < 0$ the diffusion coefficient of the substance is D_1, and at $x > 0$ equal to D_2. Let us assume that at $t = 0$ the diffusing substance is absent in the investigated region, and at the boundary $x = -l$ a uniform concentration of the diffusing substance is maintained from that instant (diffusion from a steady source). The concentration distribution $C_1(x, t)$ at $-l < x < 0$ and $C_2(x,t)$ at $x > 0$ is found by solving the differential equations (2.70) and (2.71) under conditions of conjugation (2.73) and initial and boundary conditions

$$\begin{aligned} t = 0, \quad &-l < x < 0, \quad C_1 = 0, \\ &0 < x < \infty, \quad C_2 = 0, \\ x = -l, \quad t > 0, \quad &C_1 = C_0. \end{aligned} \tag{2.78}$$

The solution of the formulated problem is found by use of the Laplace transform, and it may be expressed [2] in the form of

$$C_1(x, t) = C_0 \sum_{n=0}^{\infty} \alpha^n \left\{ \Phi^* \left[\frac{(2n+1)\, l + x}{2\sqrt{D_1 t}} \right] - \alpha \Phi^* \left[\frac{(2n+1)\, l - x}{2\sqrt{D_2 t}} \right] \right\}, \qquad (2.79)$$

$$C_2(x, t) = \frac{2C_0}{1 + \sqrt{\frac{D_2}{D_1}}} \sum_{n=0}^{\infty} \alpha^n \Phi^* \left[\frac{(2n+1)\, l + \sqrt{\frac{D_1}{D_2}}\, x}{2\sqrt{D, t}} \right], \qquad (2.80)$$

where

$$\alpha = \frac{\sqrt{\frac{D_2}{D_1}} - 1}{\sqrt{\frac{D_2}{D_1}} + 1}; \quad \Phi^*(z) = 1 - \mathrm{erf}\, z.$$

The flow of substance in the region $-l < x < 0$ at the boundary with the region $x = -l$ is equal to

$$j_D = -D_1 \left(\frac{\partial C_1}{\partial x} \right)_{x=-l} = C_0 \sqrt{\frac{D_1}{\pi t}} \left\{ 1 + 2 \sum_{n=1}^{\infty} \alpha^n \exp \left[-\frac{n^2 l^2}{D_1 t} \right] \right\}. \qquad (2.81)$$

For large times all the exponential functions in (2.81) may be replaced by units, and we obtain the approximate value

$$j_D = C_0 \sqrt{\frac{D_1}{\pi t}} \left(1 + \frac{2\alpha}{1-\alpha} \right) = C_0 \sqrt{\frac{D_2}{\pi t}}. \qquad (2.82)$$

Solutions of more complex problems of nonstationary diffusion in compound bodies may be borrowed from corresponding solutions in the theory of thermal conductivity [2–3].

§ 13. Methods of Determining Coefficients of Diffusion in Rocks

Methods of determining diffusion coefficients of solutions and gases in rocks are based on treatment of experimental data on diffusion, which may be obtained by various methods, by means of appropriate solutions of diffusion equations. It is advisable to distinguish the different methods of determining diffusion coefficients: 1) by using stationary flow of substance through the rock, 2) by the method of time lag, and 3) by using nonstationary flow. We shall examine these methods and point out the limits of their usefulness.

1. The Method of Stationary Flow. The essence of this method is the following: a sample of rock in the form of a cylinder of height l is placed in a tube (ordinarily of glass or metal) impermeable to the diffusing substance. If one end of the cylinder is held at a concentration $C = C_1 = \text{const}$ and the other at $C = C_2 = \text{const}$, then, after some time, a stationary current of substance through the rock is established. In keeping with Eq. (2.15) the amount of substance diffusing per unit time is

$$\frac{Q}{t} = D \frac{C_1 - C_2}{l} S, \qquad (2.83)$$

where S is the cross-sectional area of the cylinder.

When the quantity Q/t is determined experimentally, the diffusion coefficient D is found from Eq. (2.83).

2. The Method of Time Lag. This method for determining the diffusion coefficient, developed by Daynes [28] and later by Barrer [29] and others [3], is based on the asymptotic soution of the equation of nonstationary diffusion (2.6) for a cylindrical rock sample. The solution of Eq. (2.6) with initial and boundary conditions

$$
\begin{aligned}
t &= 0, \quad 0 < x < l, \quad C = 0, \\
t &> 0, \quad\quad x = 0, \quad\quad C = C_1, \\
t &> 0, \quad\quad x = l, \quad\quad C = 0,
\end{aligned}
$$
(2.84)

corresponding to the absence of the substance in the rock at zero time and the maintenance of steady concentration at the ends of the cylinders at $t > 0$, has the form

$$
\frac{C(x, t)}{C_1} = \frac{l-x}{l} + \frac{2}{\pi} \sum_{n=1}^{\infty} \frac{(-1)^n}{n} \sin \frac{n\pi(l-x)}{l} \exp\left(-\frac{n^2\pi^2 Dt}{l^2}\right).
$$
(2.85)

The amount of substance collected at time t by the experimenter from the face $x = l$ of the cylindrical rock sample is

$$
Q(t) = -\int_0^t SD\left(\frac{\partial C}{\partial x}\right)_{x=l} dt.
$$
(2.86)

By differentiating Eq. (2.85) according to x, substituting the resulting expression in (2.86), and integrating according to t, we obtain

$$
Q(t) = \frac{DC_1 S}{l}\left[t - \frac{2l^2}{\pi^2 D}\left\{\left[1 - \exp\left(-\frac{\pi^2 Dt}{l^2}\right)\right] - \right.\right.
$$
$$
\left.\left. -\frac{1}{4}\left[1 - \exp\left(-\frac{4\pi^2 Dt}{l^2}\right)\right] + \frac{1}{9}\left[1 - \exp\left(-\frac{9\pi^2 Dt}{l^2}\right)\right] - \cdots\right\}\right].
$$
(2.87)

The graph of the function $Q(t)$ is shown schematically in Fig. 4, from which we see that, beginning at some time t', the function $Q(t)$ becomes linear. This means that a stationary flow of substance through the rock has been established. For the time $t > t_1$ ($t \to \infty$) Eq. (2.87) is written in the form

$$
Q(t) = C_1 \frac{DS}{l}\left[t - \frac{2l^2}{\pi^2 D}\left(1 - \frac{1}{4} + \frac{1}{9} - \frac{1}{16} + \cdots\right)\right] = C_1 \frac{DS}{l}\left[t - \frac{l^2}{6D}\right],
$$
(2.88)

since the sum of the series in parentheses is equal to $\pi^2/12$. If we extend the straight line in Fig. 4 till it intersects the time axis, the intersection occurs at

$$
t_3 = \frac{l^2}{6D}.
$$
(2.89)

The value t_3 is called the "time lag" (the term is not altogether fortunate, since the actual time lag in establishing stationary flow is $t_1 > t_3$). By measuring experimentally the time lag and knowing the linear dimensions of the investigated rock, the diffusion coefficient is found from the formula

$$
D = \frac{l^2}{6t_3}.
$$
(2.90)

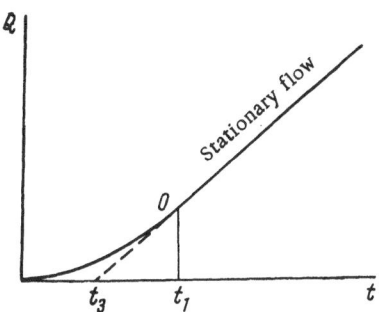

Fig. 4. Determination of time lag

This variant of the time-lag method is used for determining the diffusion coefficient of substances not interacting with the rock. When the substance is absorbed by the rock, exchanges ions, or enters into chemical reaction with substances of the rock, Eq. (2.87) does not describe diffusion of the substance in the rock.

3. The Method of Nonstationary Flow.

Determination of the diffusion coefficient of a substance under nonstationary conditions is based on measurement of the concentration distribution of the substance diffusing through the rock for some fixed time and on treatment of the results obtained by means of appropriate solutions of the diffusion equation (2.6). Let us consider the method, used in a number of works (16-20), of determining the diffusion coefficient in a solution on the basis of diffusion out of a thin layer.

If we place a thin layer of the substance to be investigated, such as salt, at the bottom of a long glass or metal tube and fill the tube with moist rock, diffusion of the substance in the rock is described, as a first approximation, by Eq. (2.40) (we assume that the x axis is directed along the tube). We place the origin at the bottom and then, taking the logarithms of both sides of Eq. (2.40), we obtain

$$\ln C = \ln \frac{Q}{2\sqrt{\pi D t}} - \frac{x^2}{4Dt} = A - Bx^2, \tag{2.91}$$

where A and B are constant values for a fixed time.

Thus, by determining the concentration distribution $C(x,t)$ of the substance at a fixed time along the length of the tube (see[16-20], for example), it is possible from the slope of the straight line $\log C(x, t)$ in the function of (x^2) to determine the diffusion coefficient D. The relations of (2.91) are fulfilled if the diffusing substance does not interact with the rock. Determination of the diffusion coefficient of substances being adsorbed, by using the thin-layer method, therefore requires a different mathematical treatment of the measurements. It should be noted, however, that in a number of papers [17, 19, and 20] the investigated method was applied not altogether systematically, since the layer of substance from which diffusion occurred was not thin.

Let us examine the nonstationary-flow method used by P. L. Antonov for determining the diffusion coefficient of gases in rocks (the method of partial saturation according to the author's terminology [31]). A cylindrical sample, sealed on the sides, in which no diffusing substance is present at the beginning of the experiment, is placed in a substance (solution or gas) with steady concentration C_0. The amount of substance adsorbed by the sample is found by solving the equation of one-dimensional diffusion (2.4) with the initial condition $C = 0$ at $t = 0$ everywhere within the cylinder and with the boundary condition $C = \beta C_0$ at $z = 0$ and $z = l$ for any $t > 0$ (the z axis is directed along the axis of the cylinder, l is the height of the cylinder, and β is the distribution factor of dissolved substance between the sample and the solution). The following expression is then obtained for the amount of substance Q adsorbed by the rock sample with a height l and radius r_0 at time t [31].

$$Q = \beta C_0 \pi r_0^2 l \left[1 - \frac{8}{\pi^2} \sum_{n=1}^{\infty} \frac{1}{(2n-1)^2} \exp\left(-\frac{(2n-1)^2 \pi^2 Dt}{4l^2} \right) \right]. \tag{2.92}$$

Figure 5 shows the dependence of the amount of adsorbed substance (in units of $Q/\beta C_0 \pi r_0^2$) on the height of the cylinder l for different values of the parameter Dt. At low values of l the dependence is linear, corresponding to complete saturation the sample. Therefore, the amount of adsorbed substance in this region increases proportionally to l. For the linear segment of Eq. (2.92) it follows that

$$Q = \beta C_0 \pi r_0^2 l, \tag{2.93}$$

whence

$$\beta = \frac{Q}{C_0 \pi r_0^2 l} = \frac{\tan \alpha}{\pi r_0^2 C_0}, \tag{2.94}$$

where α is the slope of the linear segment of the curve in Fig. 5.

At large values of l, increase in Q becomes slower, and, beginning with some height (depending on Dt), the amount of adsorbed substance ceases to depend on the height of the sample. For the horizontal segment of the curve $Q = f(l)$, Eq. (2.92) assumes the form

$$Q_\infty = 4\beta C_0 \pi r_0^2 \sqrt{\frac{Dt}{\pi}}, \tag{2.95}$$

whence

$$D = \frac{\pi}{16} \frac{1}{t} \left(\frac{Q_\infty}{\tan \alpha}\right)^2, \tag{2.96}$$

where Q_∞ is the amount of substance adsorbed by the sample as $l \to \infty$.

By determining experimentally the amount of substance diffused into cylindrical rock samples of different heights at a certain time, and by plotting a graph similar to that of Fig. 5, we may determine the diffusion coefficient D and the distribution coefficient β in accordance with Eqs. (2.94) and (2.96).

In practice it is most important to know how to determine the diffusion coefficient of substances adsorbed in rocks. This question will be considered in Chap. 4. We note that the diffusion coefficient may also be determined from data on studies of the kinetics of sorption and ion exchange from a current (see Chap. 5).

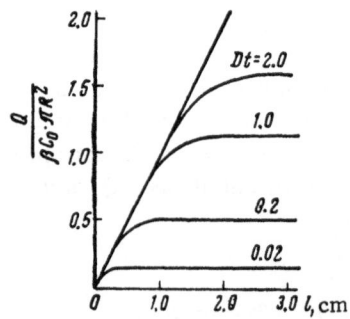

Fig. 5. Dependence of the amount of substance adsorbed by a cylindrical rock sample on the height of the cylinder.

§ 14. Description of the Diffusion of Salt in Rocks

The rules of diffusion of salt in rocks are determined by the following basic factors: 1) moisture content, 2) adsorption, ion exchange, and chemical reaction of the diffusing substance in the rock, and 3) the presence of electrical charges on the diffusing particles.

At present we have no theory giving quantitative consideration of the effect of moisture content on diffusion rate. Experimental data on diffusion in wet rocks, which will be considered below (see § 16), permit us to make only qualitative conclusions. A theory of diffusion taking into account adsorption, ion exchange, and chemical reaction between the diffusing substances and the rock will be developed in Chap. 6.

If the relative moisture content of a rock is 100% (all the pores are filled with water) and if there is no interaction between the dissolved substances and the rock, the diffusion of salt in a rock obeys the same laws as the diffusion of salt in free space.

The electrical charge of diffusing ions has a substantial effect on the diffusion of salt in water. Let us designate by C^+ and C^- the concentrations of cations and anions of salt, having a charge of Z^+ and Z^- respectively. The cations and anions differ in mobility. Let D^+ and D^- represent the diffusion coefficients of the cations and anions. The more mobile ions diffuse more rapidly, as a result of which a space charge arises, retarding the fast and accelerating the slow ions. An equality of currents is thus established. Two forces act on the ions during diffusion (in the meaning used in nonequilibrium thermodynamics [32]): the concentration gradient and the electrical field of strength E, arising as a result of diffusion. Equations of material balance for both kinds of ions are written in accordance with the Nernst−Planck equation [33] in the form

$$\frac{\partial C^+}{\partial t} = D^+ \Delta C^+ + \frac{F}{RT} D^+ Z^+ \, \mathrm{div}\,(C^+ E), \qquad (2.97)$$

$$\frac{\partial C^-}{\partial t} = D^- \Delta C^- - \frac{F}{RT} D^- Z^- \, \mathrm{div}\,(C^- E), \qquad (2.98)$$

where F is the Faraday number, T the absolute temperature, and R the universal gas constant.

Equations (2.97) and (2.98) differ from Eq. (2.6) by having a term on the right side accounting for the transfer of ions in an electrical field.

A solution must be electrically neutral at any instant of time. Therefore, during the diffusion of ions, the supplemental condition of electrical neutrality must be observed. It may be written in the form

$$Z^+ C^+ - Z^- C^- = 0. \qquad (2.99)$$

Expressing the concentrations of ions C^+ and C^- by the concentration of the electrolyte C

$$C^- = Z^+ C; \quad C^+ = Z^- C \qquad (2.100)$$

and taking into account Eq. (2.99), in place of Eqs. (2.97) and (2.98) we obtain

$$\frac{\partial C}{\partial t} = D \Delta C, \qquad (2.101)$$

where

$$D = \frac{D^+ D^- (Z^+ + Z^-)}{D^+ Z^+ + D^- Z^-}. \qquad (2.102)$$

From Eqs. (2.101) and (2.102) it follows that the diffusion of an electrolyte may be described by Fick's laws, but the diffusion coefficient becomes an effective value depending on individual diffusion coefficients of ions and their charges. Equations (2.101) and (2.102) show also that we may determine experimentally only the effective diffusion coefficient D, not the diffusion coefficients D^+ and D^- of the individual ions.

§15. Experimental Data on the Diffusion of Salt in Rocks

Some of the first to study the diffusion of salt in soils were Wollny [34] and Müntz and Gandechon [35]. To investigate the distribution of moisture during diffusion of a salt solution in soil, Müntz and Gandechon set up the following experiment [35]. A box $300 \times 300 \times 150$ mm was filled with sandy soil having a moisture content of 3.2%. Samples of crystalline sodium nitrate, 0.5 g, were placed at each corner of the box, at different distances from the edges and at a depth of 1 cm in the soil. The box was covered with a glass plate. On the following day wet spots averaging 1 cm in diameter appeared on the surface of the soil above the sites where the salt samples had been placed. The diameter of the spots gradually increased, and at the end of the eighth day reached 30-40 mm. The moisture content in the zones between spots declined to 2.6%, but in the ares of the spots it increased to 6.3%.

On the basis of several similar experiments made with sodium nitrate and potassium chloride, the authors stated that after dissolving in the soil moisture the salt was not spread throughout the entire volume of the soil. The soil after a time was divided into two sharply contrasting zones: 1) a moist zone with its dissolved salt, attracting water from the surrounding soil, and 2) a dry zone, gradually losing its moisture through migration to the salt solution but not receiving salt from the regions of the spots.

The method of Müntz and Gandechon was soon used by Malpo and Lefort [36]. These investigators observed the diffusion of calcium and sodium nitrates in soil and sand. Results of their experiments confirmed the data of Müntz and Gandechon. McCool and Wheeting [35] published results on the study of diffusion of sodium chloride and calcium chloride solutions in soil. Later investigations of the diffusion of salt solutions in soil were continued by Wheeting [38-39].

One of the first domestic investigators in the field of diffusion of solutions in soils was Shoshin [40]. He observed the diffusion of solutions of sodium, calcium, potassium, and barium chlorides, sodium nitate, ammonium and potassium sulfate and superphosphate in leached Voronezh chernozem and sandy chernozem.

Shoshin concluded that the solution of sodium nitrate diffused most energetically and that the superphosphate diffused with least energy. After three months the latter had moved only 7.5 cm. In particular, calcium chloride diffused more rapidly than barium chloride. A decline in rate of diffusion was noted when the moisture content of the soil was lowered. Diffusion of sodium chloride solution in the leached chernozem was appreciably slowed when the moisture content fell below 10%. Sodium chloride in the leached chernozem drew water from the surrounding soil up to a certain limiting distance.

Lebedev [41] studied the diffusion of lithium chloride in soil samples from horizon A of degraded chernozem. In its mechanical composition this soil was clayey. Lebedev's method was to place two moist layers of soil together, one of which contained lithium chloride at the time they were joined.

The technical procedure in its general outline was as follows. The soil was placed in a can of galvanized iron 10 cm tall and 4.5 cm in diameter. The can was filled with two layers of soil, each 5 cm thick. The soil containing the lithium chloride was placed in the lower part of the can. The upper part of the can had five slits, through which, at the end of the experiment, a special plate-like layer of soil was cut into small individual cylinders 1 cm tall. The chlorine ion concentration and the moisture content were measured for each sample. Sealing was accomplished by using paraffin. The experiments were carried out at constant temperature.

Lebedev concluded that salts do not diffuse throgh soils if the moisture content of the soil is below its maximum hygroscopicity. With sufficient moisture, the salt diffuses toward low concentrations, and the direction of the diffusing current may or may not coincide with the direction of water movement.

Diffusion experiments in moist quartz sand have been described in the paper of Polynov and Bystrov [42]. The solutions contained two different salts. The experiments were performed in glass cylinders 100 cm high and 10 cm in diameter. Samples were analyzed layer by layer for their contents of anions and moisture. The authors concluded that the chlorine ion diffuses more rapidly than the sulfate ion.

In 1935, two experimental works of V. Chernov and M. Gilis were published in a volume of collected papers. These contained different viewpoints on the problem of studying the diffusion of solutions in soil.

In the paper of Chernov [43], the method of diffusion from a thin layer, described above, was used for determining the diffusion coefficients for NO_3^- and Cl^- anions. Diffusion experiments were carried out by Chernov in cylindrical tubes of galvanized iron 5 cm in diameter and 20 cm tall. Each tube had 30 slits on one side for taking soil samples.

At the beginning of the experiment, a weighed sample of salt ($NaNO_3$, $NaCl$) was placed on the bottom of the paraffin-coated tube. Above this was added moist sand (chernozem and podsol loam). The tube was then closed with a stopper at the top and coated with paraffin, and then left for a certain time in a vertical position at constant temperature. After the assigned time the soil layers were removed and analyzed for their content of Cl^- and NO_3^-. The diffusion coefficient was determined from the dependence of the logarithm of ion concentration on the square of the distance to the bottom of the tube [cf. Eq. (2.91)]. The diffusion coefficient of NO_3^- in chernozem proved to be 0.285 cm^2/day and 0.282 cm^2/day for durations of 9 and 25 days respectively and 0.252 cm^2/day and 0.237 cm^2/day for durations of 45 and 65 days. Similar results were obtained for the podsol loam: 0.302, 0.309, 0.242, 0.246 cm^2/day for durations of 13, 26, 54, and 74 days. If the diffusion coefficients computed by Fick's law changes with time, then, consequently, diffusion is not described by this law. Chernov believed, however, that diffusion of the ions NO_3^- and Cl^- conforms to Fick's law, holding that the variability of the diffusion coefficient is associated with change in air temperature in the laboratory (because of weather changes).

In the paper of Gilis [16] results are given for experiments on diffusion of solutions of superphosphate, ammophos (ammonium phosphate fertilizer), and monocalcium and mono ammonium phosphates in soil. The method and technical procedure of the experiments were identical to those described for the experiments of Cherkov.

However, the choice of diffusing substances compelled Gilis to turn his attention to the effect of adsorption on diffusion. In considering the results of his experiments, he wrote: "It is impossible to consider the data we obtained on movements of the fertilizer we investigated to be the result solely of diffusion, since the phenomenon of adsorption plays a role of no little importance." And, further: "Diffusion in soil is accompanied by adsorption, differing for different fertilizers. This means that we cannot directly use mathematical formulas as was done by Chernov for the adsorption of nitrates by soil."

On the basis of his experiments, Gilis concluded that: "The diffusion rate is determined not only by the amount of moisture but also by the nature of interaction between the introduced fertilizer and the soil adsorbent complex. Phosphoric acid, forming relatively insoluble compounds, ceases movement not far from the site where the fertilizer was placed. Ammoniacal compounds are more soluble and therefore, in ammonium sulfate-phosphate, the ammonium ion moves away more rapidly than the ion of phosphoric acid."

In a second paper by Chernov [17], also devoted to a study of diffusion of solutions in soil, the author gives results of observations made by his earlier method on the diffusion of Cl^-, NO_3^-, CNS^-, SO_4^{-2}, and PO_4^{-3}. In discussing his experimental results, Chernov not only followed the opinion of Gilis concerning the failure to conform to Fick's laws, but he suggested that, when the moisture content is insignificant, the movement of water accompanying the diffusion also fails to obey Fick's laws.

Nevertheless, Chernov inconsistently maintained that in a number of cases, such as during the diffusion of Cl^-, NO_3^-, and CNS^- in soil, adsorption and the movement of water in the soil were insignificant and could be neglected. For these cases, Fick's law would be fulfilled, and the experimental results supposedly could be subjected to mathematical analysis on the basis of Fick's laws.

It is also pointed out in the papers of Komarova [18] and Dolgov [19] that the diffusion of solutions in soil is a complex process not conforming to Fick's law.

In Komarova's experiments, a method was used that had already been used by Chernov. Komarova carried out her experiments on samples of sand, chernozem, and Ca-rich chernozem, moistened at the beginning of the experiment to 60% of the total moisture capacity. The sources were weighed samples of dry salts: KNO_3, KH_2PO_4, $CaH_4P_2O_8$, and mixtures of $KNO_3 + KH_2PO_4$ and $KNO_3 + CaH_4P_2O_8$.

According to Komarova's data, during the diffusion of a KNO_3 solution through quartz sand, the diffusion coefficient after 15 days was found to be 0.532 cm^2/day; after 32 days it was 0.357 cm^2/day, and after 50 days 0.295 cm^2/day. Similar lessening of the diffusion rate was also observed in most of the other experiments.

Dolgov and Kamenova [19] carried out a number of diffusion experiments using Lebedev's method, working with solutions of chloride and sulfate salts in loamy soils. These investigators tried to reach a balance between the amount of solution expended and that amount detected in the soil after completion of the diffusion experiment. On the basis of their results, they stated that, during the joint diffusion of chlorine ions and sulfate ions through soil, the first move faster the second slower than the same kind of ions when they alone are diffusing. At the same time, the investigators tried to relate ion-exchange processes to diffusion in soil and to examine the dependence of the type of adsorption of salt on the moisture content of the soil.

Bouldin [44] investigated one-dimensional diffusion of the phosphates $Ca(H_2PO_4)_2$ and KH_2PO_4 in water-saturated soils by means of radioactive tracers using P^{32}. He found that the distribution curve of radioactive P^{32} along a core sample of silty loam is complex. It cannot be defined by Fick's diffusion laws. The results obtained are explained by precipitation of $Ca(H_2PO_4)_2$ on change of pH of the solution. The latter change was due to change in ion concentration along the soil column because of diffusion. However, the author did not discuss the effect of adsorption of phosphates by soils on the diffusion picture.

Dakshinamurti [45] studied the effect of the heterogeneity of the medium on the diffusion rate of the Br^- ion in kaolin, bentonite, and soils. From a comparison of the measured diffusion coefficients with the diffusion coefficients of Br^- in a free solution, the tortuosity factor η was calculated with consideration of Eq. (2.62). This coefficient proved to be approximately 0.7.

Lahav[24] measured the coefficient of self diffusion of Ca^{45} in the solid phase. A solution of radioactive Ca^{45} was brought in contact with a suspension containing particles of the minerals calcite, dolomite, and $CaCO_3$ of a definite size. After different lapsed times the suspension was centrifuged, and the activity of the solution was measured. According to the decrease in activity it was possible to determine the amount of Ca^{45} that was exchanged with Ca on the surface of the mineral and was diffused into the crystal. In treating the experimental data the author took into account the adsorption of Ca^{45} by the solid phase. The diffusion coefficient of Ca^{45} in the investigated minerals is the same for both, on the order of $\overline{D} \approx 10^{-20}$ cm^2/sec.

Thus, the roles of adsorption and ion exchange during diffusion of salt have been shown in a number of the papers examined above. A theory describing diffusion with consideration of adsorption and ion exchange, and also accounting for experimental investigations in which an attempt has been made to interpret results from the viewpoint of this theory, will be considered in Chap. 6.

§16. The Effect of Moisture Content on the Diffusion of Salt in Rocks

The diffusion of salt in rocks of different moisture contents has been studied by a number of workers [20, 22, 40, 41, 46]. It has been ascertained that with decrease in moisture content in a rock the diffusion coefficient of salt declines. According to V. A. Priklonskii the diffusion coefficient of NaCl was found to range for kaolinitic clays from 0.42 cm^2/day at a moisture content of 39% to 0.66 cm^2/day at a moisture content of 52%; for montmorillonitic clays from 0.39 cm^2/day at 58% moisture to 0.66 cm^2/day at a moisture content of 114%. There is a lower limit of moisture content below which salt will not diffuse. In a paper on the study of diffusion of alkali ions (Na^+, Li^+, NH_4^+) in bentonite [46], it has been reported that the ions do not diffuse at a relative moisture content below 25%. If all the pores are filled with water, the ions diffuse in the same order they do in solution ($D_{NH_4} > D_{Na} > D_{Li}$).

These experimental data show that the diffusion of salt in rocks takes place through interaggregate and intra-aggregate pores when the pores are partially or completely filled with water. When the moisture content of the rock declines, the volume filled with water in which the diffusion of the dissolved substances takes place declines. Consequently, the diffusion current of the substance diminishes, in accordance with Eq. (2.1) an amount equivalent to the decrease in the diffusion coefficient.

If the rock is not saturated with moisture, diffusion is accompanied by osmosis. The osmotic transfer of water may be considered diffusion of water molecules, since it is directed toward diminishing concentrations of water. Thus, the osmotic transfer of water takes place in a direction opposite to the diffusion of salt, and this leads to a diminution of the concentration gradient of dissolved substance and, consequently, to a decrease in diffusion rate.

In a number of papers [47-49], one-dimensional flow of water in soils was described by means of a diffusion equation of the type

$$\frac{\partial \theta}{\partial t} = \frac{\partial}{\partial x} \left[D(\theta) \frac{\partial \theta}{\partial x} \right], \tag{2.103}$$

where θ is the content of water per cm^3 of porous rock, and D is the diffusion coefficient of water, depending on θ.

The dependence of the diffusion coefficient on the moisture content may be taken to be either linear,

$$D(\theta) = \alpha\theta + \beta, \tag{2.104}$$

or exponential

$$D(\theta) = \alpha e^{3\theta}, \qquad (2.105)$$

where $\alpha > 0$, $\beta > 0$ are parameters characteristic of a given soil.

Solution of the systems of equations (2.103) and (2.104) or (2.104) and (2.105) presents considerable difficulty. Scott [49] offered solutions of these systems for the diffusion of water in a dry sample in the form of dimensionless graphs, obtained by means of an electronic computer.

The experimental investigation of the diffusion of water in soils has been carried on, in particular, by Gardner [50] and Dutt [51]. The discovered dependence of the diffusion coefficient on water content for different soils shows that the diffusion coefficient increases differently for different soils, with diminution of moisture content. This is in agreement with Eqs. (2.104) and (2.105). However, on the basis of the results we have obtained, we cannot decide unequivocally whether or not the dependent relations (2.104) and (2.105) are fulfilled.

Let us find a system of equations that will describe one-dimensional diffusion of salts, taking osmosis into account. Should osmotic transfer be lacking, the diffusion of salt would be described by the equation

$$\frac{\partial C^*}{\partial t} = D \frac{\partial^2 C^*}{\partial x^2}, \qquad (2.106)$$

where C^* is the concentration of salt in the absence of osmosis, and D is the diffusion coefficient of the salt.

Let us use θ_0 to designate in cm^3 the moisture content per cm^3 of rock at initial time $t = 0$. If there is no osmotic transfer of water, then

$$C^* = \frac{Q}{\theta_0 V}, \qquad (2.107)$$

where Q is the total amount of dissolved substance in volume V of rock.

Because of osmosis, however, $\theta(t) \neq \theta_0 (t > 0)$, but it varies with time and space in accordance with Eq. (2.103). As a consequence, the concentration of diffusing substance at time t becomes

$$C = \frac{Q}{\theta V}. \qquad (2.108)$$

From Eqs. (2.107) and (2.108) it follows that

$$C^* = C \frac{\theta(x, t)}{\theta_0}. \qquad (2.109)$$

By substituting the value C^* from Eq. (2.109) into Eq. (2.106) we obtain the following differential equation describing the diffusion of salt, taking into account the osmotic transfer of water:

$$\frac{\partial}{\partial t} \left(C \theta \right) = D \frac{\partial^2 (C\theta)}{\partial x^2}. \qquad (2.110)$$

The system of equations (2.103) and (2.110) describes the diffusion of salt for certain initial and boundary conditions, taking into account the osmotic transfer of water. The initial and boundary conditions must give the distribution of moisture and dissolved substance at the initial moment, $t = 0$, and at the boundaries of the medium. The solution of the system of equations (2.103) and (2.110), accounting for the initial and boundary conditions, leads to the desired functions $C = C(x,t)$ and $\theta = \theta(x, t)$.

Experimental investigation of one-dimensional diffusion of NaCl in clays in which osmotic transfer of water takes place (natural incomplete saturation of clays) has been made by N. P. Zatenatskaya [22]. Experiments were carried out in tubes with slits for removing samples for analysis after termination of the diffusion experiment. The tubes were filled with samples of clay rocks and muds with natural moisture contents and undisturbed structure. A layer of dry salt NaCl ("salt screen") was placed in the upper part of each tube. After the lapse of a set time, samples were analyzed layer by layer for their contents of Cl ion, and the diffusion coefficient of the salt was computed from Eq. (2.91).

The coefficient of diffusion is not a constant value. It depends on the distance to the "salt screen." Thus, $D \approx 0.2{-}0.56$ cm^2/day near the "salt screen" (1–3 cm) but is considerably greater, $D \approx 0.46{-}0.72$ cm^2/day at a distance of 4–8 cm. Investigation has shown that the moisture content of muds near the "salt screen" is appreciably lowered (15–20%) from the initial content. Far from the "salt screen" the moisture content approaches the values of natural moisture content. Similar results were obtained for clays: the diffusion coefficient D was found to increase from about 0.22–0.24 cm^2/day near the top of the column to about 0.27–0.55 cm^2/day near the bottom. The moisture content of the rock near the "salt screen" was also found to be lowered (4–7%), but the dry salt became moist and was partially removed in solution. The observed pattern may be explained by osmotic transfer of water, taking place in a direction opposite to the diffusion of NaCl. As a consequence, the concentration gradient of the salt in solution near the "salt screen" becomes flatter and diffusion is slowed down. Figure 6 illustrates the results of the experiments.

It was also found that, with decrease in moisture content of the rock, the change in diffusion rate with distance from the "salt screen" became less because of diminished osmotic transfer of water. In highly compacted clays, in which the content of free water is insignificant, and strongly held water occupies 30–40% of the pore space, free diffusion of NaCl takes place, uncomplicated by osmosis.

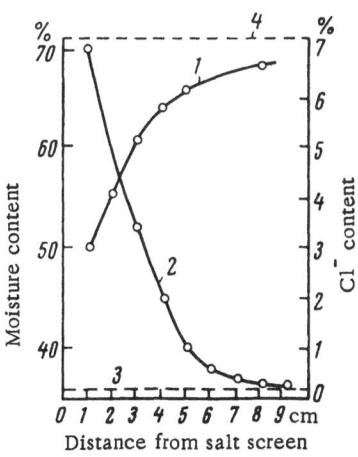

Fig. 6. Change in moisture content and salt content of mud during diffusion transfer of the Cl ion from NaCl for 5 days. 1) Moisture content; 2) salt content; 3) content of Cl before the experiment; 4) natural moisture content.

§ 17. Experimental Data on the Diffusion of Gases in Rocks

Studies on the diffusion of gases in rocks [15, 31, 52–56] have been undertaken chiefly in connection with problems of "gas surveys" [31, 52] and the aeration of soils [15, 53, 56].

In [31, 52] the diffusion of hydrocarbons in clays and sands is discussed. For describing diffusion, Antonov uses the adsorption constant (distribution coefficient) proposed by Wroblewski [57], which Antonov calls the gas capacity β and which is determined from the formula

$$\beta = \frac{Q}{V C_0}, \qquad (2.111)$$

where Q is the amount of gas occurring in volume V of the rock that is in contact with the gas of concentration C_0.

The value of β for small concentrations of diffusing substances was represented by the author in the form

$$\beta = (n_0 - n_1) + n_1 \gamma + (1 - n_0) K, \qquad (2.112)$$

where n_0 is the porosity of the given rock, computed as that part of the volume occupied by pores, n the part of the volume occupied by moisture; γ the solubility coefficient of the gas in water, and K the adsorption coefficient.

The diffusion of methane was studied by the method of "stationary flow." A formula similar to (2.83), in which the gas capacity of the rock was accounted for in accordance with (2.111) and (2.112), was used for processing measurements made to determine the diffusion coefficients. By diffusion coefficient the author has in mind some effective value, considering all forms of gas transfer in the rock. The results obtained for rock samples with natural moisture contents at a temperature of 20°C are shown in Table 1.

Measurement of the diffusion coefficients of methane and other hydrocarbons in clays and sandstones have also been made by the method of nonstationary flow. Rock samples in the form of cylinders of identical radii, but differing in length, were placed in a sealed vessel in which the gas concentration was kept constant. After the lapse of a definite time, samples were taken and the amount of absorbed gas was measured by the desorption method. In accordance with Eqs. (2.94) and (2.96) the values of D and β were determined from the curve of amount of adsorbed gas versus height of cylinder (a theoretical form of this curve is shown in Fig. 5). The values obtained for D and β for the diffusion of methane in clays having a moisture content of 20-25% (by weight) are given in Table 2.

The diffusion of various hydrocarbons was studied in clay of the Kudinovskii pit. Results are shown in Table 3.

From Table 1 it may be seen that the diffusion parameters D and β depend little on the kind of clay, but they range between wide limits for different kinds of rock (clay, shale, sandstone) (see Table 2). From the data in Table 3 it follows that the diffusion coefficients of hydrocarbon gases are inversely proportional to the square root of their molecular weights. These results are in agreement with theory (see §11).

Antonov measured the diffusion coefficients of hexane in mixtures of quartz sand and clay. He found that the diffusion coefficient diminished with increase in the sand content.

We should note that Antonov used Fick's laws to describe the diffusion of gases in the adsorbent rocks, but these laws do not take account of adsorption of the diffusing substance. Fick's laws (see Chap. 6) may be used for describing diffusion of adsorbed substance when the adsorption isotherm is linear (which corresponds to experimental results [31, 52]). But in this case the diffusion coefficient loses its ordinary meaning and becomes an effective value, depending on the adsorption coefficient.

A number of authors [53-55] have studied the diffusion of gases (carbon disulfide, acetone, water vapor, etc.) through different air-dried and moist porous materials (sand, kaolin, soil, mica, etc.) in order to ascertain the connection between the diffusion coefficients D and D_0 of gases in porous materials and in air. Figure 7 shows the dependence of D/D_0 on free porosity (all pores filled

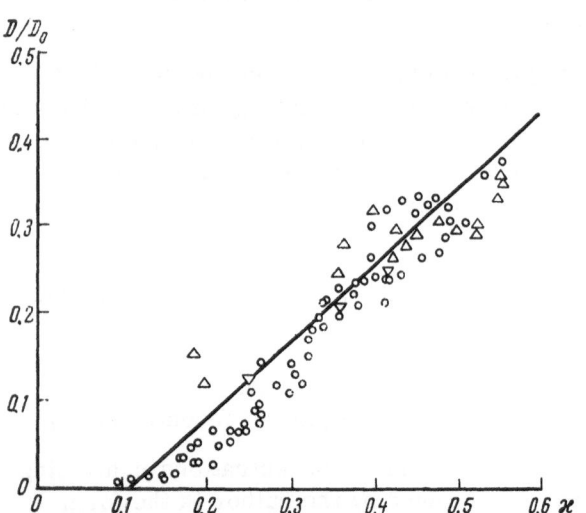

Fig. 7. Dependence of D/D_0 on free porosity ($\varkappa$) for gases according to data of Van Bavel ($\triangle$) Penman ($\triangledown$) and Taylor ($\bigcirc$).

TABLE 1. Parameters of Diffusion through Clay Rocks

Characteristics of the clay, depth, and locality	cm^2/sec	$\beta \cdot 10^{-2}$
Clay of the Tula Formation, 36-42	4.2	1.9
Mesozoic clay, Tula, 18 m	3.1	4.0
Quaternary, slightly sandy loam, Tula, 4 m. . .	2.7	4.3
Dark gray clay, from a sooty-carbonaceous coal-bearing sequence, Tula, 26 m	3.1	3.25
Dark gray clay with bituminous bedding planes, coal-bearing sequence, Tula.	3.1	2.7
Clay from the vicinity of Varenikovskaya, Shugo Volcano, 2 m	4.8	1.55
Gray-blue clay, viscous, Shugo Volcano, 2 m .	5.2	1.45
Brown-green clay, swampy flood-plain of the Kuban' River, 2 m	3.1	2.0
Gray-blue clay, swampy flood-plain of the Kuban' River, sandy and containing limestone fragments, 2 m.	1.35	4.3
Gray-brown clay, highly ferruginous, swampy flood-plain of the Kuban' River, 2 m	2.7	2.9
Clay from Lok Batan (breccia of last eruption).	5.4	2.4
Another sample of same clay from Lok Batan) .	7.0	2.6
Clay of the Kudinovskii pit (average of six experiments).	3.93	1.83

TABLE 2. Parameters of Diffusion of Methane through Clay Rocks

Rock	D, cm^2/sec	β
Plastic clayey rocks (Tertiary and Mesozoic deposits from the regions of Stavropol', Tula Oblast, Podmoskov'e, Krasnodar, Turkmenia, Baku, and elsewhere).	$\sim 10^{-6}$	$\sim 10^{-2}$
Saturated sands.	$\sim 10^{-6}$	$\sim 10^{-2}$
		$\sim 10^{-1}$
Clayey rocks with bedded structure (Tertiary deposits of Stavropol').	$10^{-3} - 10^{-4}$	$\sim 10^{-2}$
Sandstones and siltstones with clay cement (tertiary deposits of Stavropol')	$\sim 10^{-3}$	$\sim 10^{-2}$
Clayey rocks from Tertiary deposits in Turkmenia	$\sim 10^{-5}$	$\sim 10^{-2}$
Cemented sandstones, siltstones, mudstones, and limestones (Carboniferous rocks of the Ukhta region, Komi ASSR, and of the Kuibyshev region).	$\sim 10^{-7}$	$\sim 10^{-1}$
		$\sim 10^{-2}$
Shales, silty cemented clays (Carboniferous of the Kuibyshev region)	$\sim 10^{-8}$	$\sim 10^{-1}$
Mudstones, quartz sandstones and siltstones, and limestones (Devonian of the Kuibyshev region, and Carboniferous of the upper Pechora region)	$\sim 10^{-9}$	

TABLE 3

Hydrocarbon	Molecular weight	$D \cdot 10^6$ cm^2/sec	$D \sqrt{M} \cdot 10^6$	$\beta \cdot 10$
Methane	16	3.95	15.8	1.83
Ethane	30	3.50	19.1	2.25
Propane	44	2.67	17.8	1.66
Butane	58	2.48	18.9	1.68
Hexane	86	1.55	14.5	1.70

TABLE 4. Values of the Diffusion Coefficient of CO_2 Depending on the Structural State and Density of Ordinary Clayey Chernozem (bulk weight of aggregates, 1.60 g/cm³; specific gravity of solid phase of soil, 2.59)

Soil	Density, g/cm³	Moisture content, % of volume	Porosity, %		Free pores, %	$D_0 (CO_2-air)$ at temperature of experiment	D	$\frac{D}{D_0}$
			Total	Inter-aggregate				
Original soil	0.834	6.1	67.8	47.8	61.7	0.159	0.0499	0.312
	0.904	6.6	65.1	43.5	58.5	0.159	0.0461	0.290
	0.986	7.2	62.0	38.4	54.8	0.160	0.0410	0.256
Dust (particles less than 0.25 mm)	0.846	5.5	67.3	—	61.8	0.159	0.0474	0.298
	0.948	6.2	63.4	—	57.2	0.159	0.0434	0.274
	1.025	6.7	60.4	—	53.7	0.157	0.0426	0.271
	1.120	7.3	56.8	—	49.5	0.161	0.0366	0.227
	1.136	7.4	56.2	—	48.8	0.157	0.0384	0.245
Aggregates (0.25-0.5 mm)	0.791	5.8	69.5	50.5	63.7	0.163	0.0546	0.334
	0.814	5.9	68.6	49.2	62.7	0.163	0.0528	0.323
	0.914	6.7	64.7	42.8	58.0	0.163	0.0465	0.285
	0.930	6.8	64.1	41.8	57.3	0.163	0.0451	0.276
	0.958	7.0	63.0	40.2	56.0	0.163	0.0440	0.270
	0.967	7.0	62.7	39.6	55.7	0.163	0.0432	0.265
Aggregates (2.5-3.0 mm)	0.718	5.2	72.3	55.2	67.1	0.157	0.0547	0.349
	0.743	4.5	71.3	53.5	66.8	0.161	0.0566	0.352
	0.745	4.5	71.2	53.4	66.7	0.159	0.0542	0.342
	0.758	4.5	70.7	52.7	66.2	0.160	0.0546	0.342
	0.783	4.9	69.7	51.0	64.8	0.160	0.0541	0.338
	0.830	6.1	67.9	48.2	61.8	0.158	0.0500	0.316
	0.843	5.6	67.4	47.3	61.8	0.160	0.0502	0.312
	0.895	6.5	65.5	44.0	59.0	0.157	0.0465	0.296

with air are included in free porosity) according to data from a number of authors [53–55]. From these data, and also from theoretical considerations, it follows that the diffusion rate of gas is proportional to the free porosity of materials. From Fig. 7 it follows that approximately 10% of the free pore volume does not participate in diffusion, a fact that may be explained by the presence of noncommunicating and sharply constricted pores.

Poyasov [15, 56] studied the diffusion rate of CO_2 in soils and its dependence on the structural state and moisture content. The diffusion coefficient of CO_2 in soil was determined by using one of the variants of the nonstationary-flow method, on which we shall not dwell. In Table 4 we have given the values of the diffusion coefficient of CO_2 through air-dried clayey chernozem of different structural states. These data establish a connection between the diffusion coefficients

TABLE 5. Values of the Coefficient of Diffusion of CO_2 for
Several Soils in Dependence on Moisture Content

Soil	Density, g/cm³	Specific gravity	Total porosity, %	Moisture, % of volume	Free pores, %	D_0 (CO_2–air) at temperature of experiment	D	$\frac{D}{D_0}$
Ordinary clayey chernozem	0.846	2.59	67.3	6.2	61.1	0.161	0.0473	0.294
	0.846	2.59	67.3	15.3	52.0	0.161	0.0352	0.218
	0.846	2.59	67.3	23.7	43.6	0.161	0.0261	0.162
	0.846	2.59	67.3	30.6	36.7	0.161	0.0203	0.126
	0.846	2.59	67.3	30.0	37.3	0.160	0.0239	0.149
	0.846	2.59	67.3	28.7	38.6	0.160	0.0267	0.167
	0.846	2.59	67.3	36.1	31.2	0.160	0.0172	0.107
	0.846	2.59	67.3	39.6	21.7	0.160	0.0160	0.100
Dusty, ordinary clayey chernozem (Kamennaya Steppe). Particles less than 0.25 mm	0.819	2.59	68.4	4.9	63.5	0.161	0.0521	0.323
	0.819	2.59	68.4	13.7	54.7	0.160	0.0342	0.213
	0.819	2.59	68.4	22.6	45.8	0.159	0.0233	0.147
	0.819	2.59	68.4	22.1	46.3	0.159	0.0264	0.174
	0.819	2.59	68.4	28.7	39.7	0.159	0.0166	0.104
	0.819	2.59	68.4	36.0	32.4	0.159	0.0134	0.084
Turfy, slightly podsol-like, heavy, loamy soil; field sample. Horizon A_1, grass in 2nd year of use	1.46	2.70	45.9	24.5	21.4	0.159	0.0106	0.067
	1.46	2.70	45.9	22.2	23.7	0.158	0.0117	0.074
	1.46	2.70	45.9	19.0	26.9	0.157	0.0123	0.078
Turfy, podsol, heavy loamy soil. Soil crust; field sample	1.30	2.58	49.6	5.1	44.5	0.162	0.0230	0.142
	1.30	2.58	49.6	8.8	40.8	0.161	0.0182	0.117
	1.30	2.58	49.6	12.6	37.0	0.159	0.0148	0.093
	1.30	2.58	49.6	22.2	27.4	0.159	0.0076	0.048
	1.30	2.58	49.6	22.2	27.4	0.161	0.0092	0.056
Turfy, podsol, heavy, loamy soil. Field sample taken from a wet place of winter rust	1.13	2.62	56.9	50.2	6.7	0.161	0.0012	0.008
	1.13	2.62	56.9	50.2	6.7	0.158	0.0009	0.006
The same, but an air-dried sample	1.32	2.62	49.7	4.6	45.1	0.158	0.0196	0.123

D and D_0 of the gas in porous media and in air as this depends on the porosity of the rock. Poyasov, in analyzing the dependence of D/D_0 on the total, interaggregate (interparticle) and free porosities, concluded that, for strongly structured soil, diffusion rate is proportional to interaggregate porosity. However, if the structure of the soil has been disturbed (some aggregates deformed or shattered), the difference between interaggregate and intra-aggregate pores partially disappears, and diffusion takes place with the participation of the intra-aggregate pores. Within limits, diffusion takes place through all intra-aggregate pores in a structureless mass. From the data of Table 4 it follows that diffusion in air-dried soil takes place rather quickly, at a rate less than the diffusion rate in air by only a factor of 3-5.

In [15] the diffusion rate of CO_2 in soils depending on their moisture content is discussed (Table 5). It was found that the diffusion rate declines with increase in moisture content. This relation obtains because of swelling of the soil particles and closing off of the pores through which diffusion of the gas takes place. The closing off of pores is more extensive in granular structureless soils than in structured soils. Analysis of the experimental data led Poyasov [56] to draw the qualitative conclusion that "in damp and wet soils, interaggregate or, in general, noncapillary pores are of most importance in diffusion transfer."

Lastly, in considering the tortuosity factor (about which we spoke above), Poyasov came to the following conclusion concerning the relation between the diffusion coefficient in a porous medium and that in air:

$$D = \eta \, (\varkappa - \varkappa_0) \, D_0,$$ (2.113)

where $\varkappa$ is the total porosity of air-dried soil, expressed as part of the total volume, and $\varkappa_0$ is the relative volume of pores not participating in diffusion. The numerical values the tortuosity factor may assume for a system of loosely packed spheres were reported above.

LITERATURE CITED

1. Bronshtein, I. N., and Semendyaev, K. A., Handbook of Mathematics [in Russian], Izd. Fiz-Mat. Lit., Moscow (1959).
2. Carslaw, H. S., and Jaeger, J. C., Conduction of Heat in Solids, Clarendon Press, Oxford, (1947).
3. Lykov, A. A., Theory of Thermal Conductivity [in Russian], Gostekhizdat, Moscow (1948).
4. Barrer, R. M., Diffusion in and through Solids, Cambridge University Press, Cambridge (1941).
5. Boltaks, B. I., Diffusion in Semiconductors [in Russian], Izd. Fiz-Mat. Lit., Moscow (1961).
6. Gertsriken, S. D., and Dekhtyar, I. Ya., Diffusion in Metals and Alloys in the Solid Phase [in Russian], Izd. Fiz-Mat. Lit. Moscow (1960).
7. Jahnke, E., and Emde, F., Special Functions [in Russian], Izd. Nauka, Moscow (1964).
8. McKay, A. T., Proc. Phys. Soc., No. 42, p. 547 (1930).
9. Timofeev, D. P., The Kinetics of Sorption [in Russian], Izd. AN SSSR, Moscow (1962).
10. Chapman, S., and Cowling, T. G., The Mathematical Theory of Non-uniform Gases, Cambridge (1939).
11. Hirschfelder, J. O., Curtiss, C. H., and Bird, R. B., The Molecular Theory of Gases and Liquids John Wiley, New York (1954).
12. Panchenkov, G. M., Dokl. Akad. Nauk SSSR, Vol. 118, No. 4, p. 755 (1958).
13. Frenkel', Ya. I., The Kinetic Theory of Liquids [in Russian], Izd. AN SSSR (1945).
14. Glasstone, S., Laidler, K. J., and Eyring, H., The Theory of Rate Processes, McGraw Hill Book Company, New York (1941).
15. Poyasov, N. P., Collected Works on Agronomic Physics (Sb. Trudov po Agron Fiz.) [in Russian], No. 8, p. 190 (1960).
16. Gilis, M. B., Vsesoyuzn. NII. Udobrenii, Agrotekhniki i Agropochvovedeniya im. K. K. Gedroitsa, No. 7 (1935).
17. Chernov, V., Tr. Pochvennogo Inst. im. V. V. Dokuchaeva, No. 20 (1939).
18. Komarova, N., Problems of Soviet Soil Science [in Russian], No. 4 (1937).
19. Dolgov, S., and Kameneva, Z., Tr. Vsesoyuzn. NII. Udobrenii, Agrotekhniki i Agropochvovedeniya im. K. K. Gedroitsa, Fizika Pochv., No. 18 (1937).
20. Priklonskii, V. A., and Oknina, I. A., in: Questions on Engineering-Geological Studies of Clay Rocks of the USSR [in Russian], Izd. AN SSSR, p. 41 (1959).
21. Scott, E. T., and Hanks, R. I., Soil Sci., Vol. 94, No. 5, p. 314 (1962).
22. Zatenatskaya, I. P., Dokl. Akad. Nauk SSSR, Vol. 152, No. 3, p. 717 (1963).
23. Antonov, P. L., in: Geochemical Methods of Prospecting for Oil [in Russian], p. 60 (1950).
24. Lahav, N., and Bolt, G., Soil Sci., Vol. 97, p. 293 (1964).
25. Pospelov, G., Kaumanskaya, P., and Saratovkin, D., Zap. Vsesoyuzn. Mineralogich. Obshch., No. 4, p. 382 (1961).
26. Pospelov, G., and Kaumanskaya, P., Kolloid. Zh., Vol. 25, p. 215 (1963).
27. Pospelov, G., and Kaumanskaya, P., Izv. Sibirsk. Otdeleniya AN SSSR, Ser. Geologiya i Geofizika, No. 5, p. 35 (1965).
28. Daynes, H. A., Proc. Roy. Soc., Vol. A97, p. 286 (1920).
29. Barrer, R. M., Trans. Faraday Soc., Vol. 35, p. 628 (1939).

30. Deryagin, B. V., in: Methods of Investigating the Structure of Highly Dispersed and Porous Bodies [in Russian], Izd. AN SSSR (1958).
31. Antonov, P., Tr. NII Geofiz. i Geokhim. Metodov Razvedki, No. 2 (1957).
32. de Groot, S. R., and Mazur, P., Non-equilibrium Thermodynamics, North Holland Publ. Co., Amsterdam; Interscience Publishers, New York (1962).
33. Hellferich, F., Ion Exchange, McGraw-Hill, New York (1962).
34. Wollny, M., Vierteljahrsschr. der Bayrischen Landwirtschaft, Ergänzungsband, Heft 1 (1898).
35. Müntz, A., and Gaudechon, H., Annales de la Science Agronomique, Vol. 11, p. 208 (1909).
36. Malpo, L., and Lefort, G., Annales de la Science Agronomique, p. 241 (1912).
37. McCool, M., and Wheeting, L., Agron. Res., Vol. 11 (1912).
38. Wheeting, L., Soil Sci., Vol. 19, No. 4 (1925).
39. Wheeting, L., Soil Sci., Vol. 19, No. 6 (1925).
40. Shoshin, A., Izv. Donskogo Inst. Sel'sk. Khoz-va, No. 9 (1929).
41. Lebedev, A., Tr. Pochvennogo Inst. im. V. V. Dokuchaeva, No. 3-4 (1930).
42. Polynov, B., and Bystrov, S., Pochvovedenie, No. 3 (1952).
43. Chernov, V., Tr. Vsesoyuzn. NII Udobrenii, Agrotekhniki i Agropochvovedeniya im. K. K. Gedroitsa, No. 7 (1935).
44. Bouldin, D., and Black C., Soil Sci. Soc. Proc., Vol. 18, p. 225 (1954).
45. Dakshinamurti, C., Soil Sci., Vol. 88, p. 209 (1959).
46. Husted, R., and Low, P., Soil Sci., Vol. 77, p. 343 (1954).
47. Gardner, W., and Mayhugh, M., Soil Sci. Soc. Amer. Proc., Vol. 22, p. 197 (1958).
48. Klute, A., Soil Sci. Soc. Amer. Proc., Vol. 16, p. 144 (1952).
49. Scott, E., and Hanks, R., Soil Sci., Vol. 94, p. 314 (1962).
50. Gardner, W., and Miklich, F., Soil Sci., Vol. 93, p. 271 (1962).
51. Dutt, G., and Low, P., Soil Sci., Vol. 93, p. 195 (1962).
52. Antonov, P. L., in: Direct Methods of Prospecting for Oil and Gas [in Russian], Tr. Vsesoyuzn. NII Yadernoi Geofiziki i Geokhimii (1964).
53. Penman, H. J., Agric. Sci., Vol. 30, pp. 437, 570 (1940).
54. Taylor, S., Soil Sci. Soc. Amer. Proc., Vol. 14, p. 55 (1949).
55. Van Bavel, C., Soil Sci., Vol. 72, p. 33 (1951); Vol. 73, p. 91 (1952).
56. Poyasov, N., A Collection of Papers on Methods of Investigation in the Field of Soil Physics [in Russian], p. 134, Leningrad (1964).
57. Wroblewski, S., Wied. Annalen d. Physik, Vol. 8, p. 29 (1879).

CHAPTER 3

ADSORPTION AND ION EXCHANGE IN THE INTERACTION OF SOLUTIONS AND GASES WITH ROCKS

The geochemical migration of solutions and gases is normally accompanied by interaction of the substance with the rocks. The principal processes of interaction, as noted in Chap. 1, are adsorption, ion exchange, and chemical reaction. The laws of geochemical migration are determined by these processes.

In this chapter we shall examine the equilibrium of adsorption and ion exchange in the interaction of solutions and gases with rocks. The problems of equilibrium are of considerable interest, since equations of geochemical migration may be obtained by using constants characterizing the equilibrium of adsorption and ion exchange (Chaps. 4, 5, and 6).

§ 18. The Concept of Adsorption

By sorption we mean the soaking up of gaseous or liquid substances by solids or by liquids from the surrounding space. We normally distinguish between adsorption and absorption. Adsorption is the process of concentrating substance on the surface of a solid or liquid (i.e., at the gas−liquid, gas−solid, or liquid−solid interface). In absorption the substance is soaked up by the entire volume of a body. A substance that is adsorbed is called the adsorbate, and the body on which the adsorbate is attached is called the adsorbent. In our further discussions we shall consider only solids in the role of adsorbents.

The cause of adsorption lies in surface phenomena that take place at the interface between phases. They are due to forces of attraction acting on particles (atoms, molecules, ions) from the surface of the adsorbent rather than to forces acting from within the body of the adsorbent. Actually, particles within the body are acted on by attractive forces of other particles uniformly in all directions, so that the resultant forces equal zero. The attractive forces acting on a particle of the absorbent on the surface are not compensated and the resultant of these forces is not zero. It has a value that is directed perpendicular to the surface and inward toward the surface (more details concerning adsorption interactions may be found in [1-3]. Because of this, the surface possesses adsorbent properties. It attracts and holds particles of substance from the neighboring phase (liquid or gaseous).

Adsorption is greater the more interface surface there is, in the case of solids, the more of the solid surface that is exposed. If the interface surface is small, the role of surface phenomena is small, and adsorption may be neglected. Many bodies, however, (coal, silica gel, soot, soil, rock, etc.) have extensive surface area, and in considering physicochemical processes in which they are involved, adsorption cannot be neglected. The surface area per gram of adsorbent is called specific surface. For many adsorbents, especially those that contain pores (porous adsorbents), the specific surface may amount to thousands of square meters per gram.

We distinguish between physical and chemical adsorption, or chemosorption. During chemosorption, chemical reaction between adsorbate and particles of the adsorbent, with the formation of new compounds, takes place. During physical adsorption, the adsorbate tries, without changing its chemical nature, to occupy spontaneously the entire surface of the adsorbent. Inhibiting this process, opposing adsorption, is desorption, caused, like diffusion, by the tendency toward uniform distribution of substances as a consequence of thermal movement. Physical adsorption is therefore a reversible process (see Chap. 4) in the sense that it takes place in both forward and reverse directions. As a result, an adsorption equilibrium is established, in which the number of adsorbing and desorbing particles per unit time is the same. Thus, adsorption equilibrium is a dynamic equilibrium. The most important characteristic of this equilibrium is the adsorption isotherm, which shows the dependence of the amount of adsorbed substance, at constant temperature, on the equilibrium pressure or concentration of the adsorbate. The amount of adsorbed substance is expressed in grams per square centimeter of surface, or, if the surface area is unknown, in grams per gram of cubic centimeters of adsorbent. The most typical adsorption isotherm is shown in Fig. 8 (isotherm I). At high pressures (concentrations) saturation occurs, $q = q_0$, at which the entire surface area of the adsorbent is occupied by adsorbed substance. Freundlich has proposed an empirical equation to describe isotherms. For the adsorption of gases

$$q = KP^n, \qquad (3.1)$$

where P is the pressure of the gas, and K and n are constant values, determined by experiment; normally $n \approx 0.2\text{-}1$.

For the adsorption of liquid solutions

$$q = KC^n, \quad n \approx 0.2 - 0.5. \qquad (3.2)$$

Equations (3.1) and (3.2) poorly describe the initial segment of the isotherms and do not characterize saturation. They are valid for the interval of intermediate pressures and concentrations. Experimental isotherms, on the whole, are better obtained theoretically by equations that will be examined below.

§ 19. Adsorption Isotherms of Gases and Vapors on a Homogeneous Surface

Most commonly the surface of an adsorbent is inhomogeneous. However, simple theoretical rules may be obtained only for adsorbents with homogeneous surfaces, on which the concentration of "free sites" is the same for every point. An example of such a surface may be activated carbon black [3]. The process of adsorption of an individual gas may be considered formally as the following reversible chemical reaction (see Chap. 4):

$$\begin{array}{c} \text{Desorbed} \\ \text{gas} \end{array} + \begin{array}{c} \text{Free sites on} \\ \text{the adsorbent} \end{array} \underset{K_2}{\overset{K_1}{\rightleftarrows}} \begin{array}{c} \text{Adsorbed} \\ \text{gas} \end{array} \qquad (3.3)$$

where K_1 and K_2 are the constant adsorption and desorption rates, respectively.

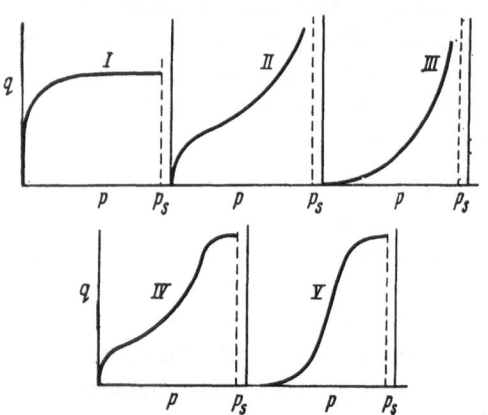

Fig. 8. Five type of adsorption isotherms (I, Langmuir; II, S-shaped; remaining types have no special names). P_s represents pressure of saturated vapor.

We shall use q_0 to designate the concentration of free sites on the adsorbent at the beginning of adsorption. We shall assume that each molecule of adsorbed gas occupies one free site. As a result of adsorption a monomolecular adsorption layer is formed on the surface (monomolecular adsorption), the thickness of which is determined by the size of the adsorbate molecules. If the concentration of adsorbed gas at time t is equal to q, the concentration of free sites at time t is equal to $q_0 - q$. In accordance with the basic postulate of chemical kinetics (Chap.4), the equation for the reaction rate (3.3) has the form

$$\frac{\partial q}{\partial t} = K_1 P (q_0 - q) - K_2 q. \tag{3.4}$$

At equilibrium the rate of the forward reaction is equal to the rate of the reverse reaction, i.e., $\partial q / \partial t = 0$. The relation between amount of adsorbed gas q and pressure at equilibrium is expressed by the following :

$$q = \frac{q_0 K P}{1 + K P} = \frac{a P}{1 + b P}, \tag{3.5}$$

$$K = \frac{K_1}{K_2}, \quad a = q_0 K, \quad b = K, \tag{3.6}$$

where $K = K_1/K_2$ is the equilibrium constant of reaction (3.3).

Equation (3.5) was first obtained by Langmuir. It is called the Langmuir adsorption equation. At low pressure (KP $\ll$ 1) we obtain

$$q = q_0 K P = K' P, \tag{3.7}$$

i.e., the adsorption is proportional to the pressure of the adsorbate. Equation (3.7) is the Henry adsorption equation, in which K' is Henry's constant, or, as we shall call it below, the adsorption coefficient.

If monomolecular adsorption takes place from a mixture of gases, we may similarly derive an expression for an adsorption isotherm of the i-th component K_i (i = 1, 2,..., n) of the mixture:

$$q_i = \frac{q_0 K_i P_i}{1 + K_1 P_1 + K_2 P_2 + \ldots + K_i P_i + \ldots + K_n P_n}, \tag{3.8}$$

where p_i (i = 1, 2, ..., n) represents the partial pressure of the i-th component and K_i (i = 1, 2, ..., n) is the equilibrium constant of reaction (3.3) written for the i substance.

Experimental adsorption isotherms of gases on many adsorbents have been obtained by Eqs. (3.5) and (3.8), in which the linear segment, in accordance with Eq. (3.7) always corresponds to low pressures. An exception is found in adsorption isotherms of vapors that begin to deviate from the theoretical dependence of (3.5) in the range of high pressures (as P approaches P_s , the saturation vapor pressure of the liquid at a given temperature). The isotherms have the forms illustrated in Fig. 8 (isotherm II). Further increase in adsorption after the monomolecular layer has formed is explained by the fact that as P approaches P_s polymolecular adsorption begins, giving rise to second, third, and more layers of adsorbed molecules. The process is terminated by volume condensation at $P = P_s$. An isotherm equation for polymolecular adsorption has been derived by Brunauer, Emmett, and Teller, generally called the BET equation. We shall not examine it here.

§ 20. Adsorption of Gases and Vapors by Porous Adsorbents

Whereas the specific surface of nonporous adsorbents may be no more than 10 m^2/g, the specific surface of porous adsorbents may reach thousands of square meters per gram [1, 3] thanks to the penetration of the entire volume of the adsorbent by fine pores. Adsorption by finely porous adsorbents has a number of peculiar features. The adsorptive forces of attraction of the walls, acting on molecules, accumulate in constricted pores, and adsorption therefore takes place more intensely in these pores than in large pores.

Because of the inhomogeneity of pores in actual adsorbents, the concentration of the adsorbate ceases being the same at every point, so that the conclusions discussed above for the Langmuir isotherm cannot be extended to the adsorption of porous adsorbents. However, experimental data are in rather good agreement with the adsorption isotherms of Langmuir, Henry, and BET [although the constants a and b do not have such clear physical significance as in (3.5)].

A characteristic feature of adsorption of vapors on finely porous adsorbents is that the isotherm has a shape similar to that for polymolecular adsorption (Fig. 8, isotherm II).

The nature of this phenomenon is different from adsorption on homogeneous surfaces, and it involves the fact that, because of the formation of concave menisci in fine pores, vapors are condensed at lower pressures than the saturation pressure of vapor for the liquid at the given temperature [1, 3]. Concave menisci form as a result of the joining of layers of adsorbed molecules appearing on the walls of fine pores because the liquid wets the walls. As shown by Thomson, the greater the curvature of a concave meniscus the lower the vapor pressure above it. Consequently, the vapor becomes saturated above the concave menisci and is condensed in fine pores at $P < P_s$. This phenomenon bears the name of capillary condensation and is explained by the rapid increase in adsorbability with pressure (Fig. 8, isotherm II).

In a number of cases better agreement with experimental data is obtained by using other adsorption isotherms. An example is found in the Dubinin–Radushkevich adsorption isotherm [4], which describes adsorption of organic substances by activated coal:

$$q = \frac{W_0}{V} e^{\dfrac{BT^2}{\beta^2 \log\left(\frac{C_s}{C}\right)^2}}, \tag{3.9}$$

where W_0 and B are the structural characteristics of the adsorbent, β is the coefficient of affinity [4], C_s and C are the vapor saturation concentration and equilibrium saturation corresponding to adsorption q, and V is the molar volume of vapor in the liquefied state.

The adsorption of gas depends also on the temperature and size of pores in the adsorbent [3, 4]. The dependence of the amount of adsorbed gas or vapor on temperature at constant pressure gives isobaric adsorption. With increase in temperature, the amount of adsorbed gas diminishes, which is characteristic of physical adsorption. The principal types of adsorption isotherms are shown in Fig. 8.

§ 21. Adsorption of Solutions on the Surface of Solids

Adsorption of solutions is subdivided into molecular adsorption of nonelectrolytes and adsorption of electrolytes. Molecular adsorption from solutions differs from adsorption of an individual gas by the fact that the solution consists of at least two components: the dissolved substance and the solvent, each of which is adsorbed. Thus, adsorption of a mixture takes place during which competition arises between molecules of the dissolved substance and the solvent for "free sites" on the adsorbent. Adsorption of the solvent is generally small, however. Experimental data show that adsorption of solutions (despite the presence of complicating circumstances) comforms for the most part to Langmuir's adsorption isotherm.

$$q = \frac{aC}{1 + bC}, \tag{3.10}$$

and for low concentrations to the Henry isotherm

$$q = KC.$$

(3.11)

The adsorption of a substance depends also on the nature of the solvent, the temperature, the size of the pores, and other properties of the system [3, 5].

The adsorption of electrolytes is very specific and ordinarily takes place when the surface of the adsorbent is charged (made up of ions or polar molecules). By virtue of electrostatic attraction, the surface adsorbs oppositely charged ions. The oppositely charged ions remain near the adsorbed ions, forming a double electrical layer [5]. The nature of the ions has the greatest effect on adsorption of electrolytes. The greater the charge on the ion the better it is adsorbed (for example, Al^{3+} is better adsorbed than Mg^{2+}, and Mg^{2+} better than Na^+). Ions of the same valence are adsorbed better the larger their radius. This is due to increase in hydration or solvation with decrease in radius [5], and this leads to decline in electrostatic interaction with the charged surface of the asdorbent, i.e., to a decrease of adsorption. According to increasing degree of adsorbing capacity, ions may be arranged in the following so-called lyotropic series [5]:

$$Li^+ < Na^+ < K^+ < Rb^+ < Cs^+$$
$$Mg^{2+} < Ca^{2+} < Sr^{2+} < Ba^{2+}$$
$$Cl^- < Br^- < NO_3^- < I^- < CNS^-.$$

§ 22. Ion Exchange

Let an electrolyte be adsorbed on the surface of a solid. When the adsorbent is placed in a solution of another electrolyte, a spontaneous process of ion exchange will take place between the adsorbent and the solution. Ion exchange continues until equilibrium is established, at which time the adsorbent and the solution contain ions in a definite quantitative ratio. Ordinarily the exchange takes place not only with ions on the surface but also with ions in the body of the material, causing dissociation of molecules of the adsorbent as a result. Solids having the capacity for body exchange of ions are called ionites [6, 7]. Schematically, an ionite might be represented [6] as consisting of a framework bound by valency forces. The framework possesses an excess charge (positive or negative) that is compensated by oppositely charged ions (counter ions). Counter ions within the framework are mobile, and when the ionite is placed in an electrolyte they may be replaced by ions from the solution. Depending on the sign of the charge of the counter ions, cationites (positive charge) and anionites (negative charge of counter ions) are distinguished. Since the ionite–solution system must be electrically neutral at any instant, during exchange, for each equivalent ion adsorbed from the solution, the ionite yields to the solution one equivalent ion with a charge of the same sign. Thus, in contrast to adsorption, ion exchange is a stoichiometric substitution of ions.

The concentration of counter ions in an ionite capable of exchanging defines the exchange capacity of the ionite. The exchange capacity of ionites is generally expressed in milligram equivalents per gram of exchanger (it does not exceed several meq/g).

Ion exchange is generally accompanied by subsidiary processes of penetration of the solvent and dissolved substance into the pores of the ionite. Adsorption of the solvent causes swelling of the ionite. Nonexchange adsorption of the electrolyte is slight. It will not be considered in future discussion.

Many authors [8–14] have established the fact that ion exchange is a reversible process, obeying the law of mass action [3]. The exchange reaction of univalent ions in a cationite may be written in the form

$$A^+ + BR \rightleftarrows B^+ + AR, \qquad (3.12)$$

where A^+ and B^+ are exchanging cations, and AR and BR are the ionic forms of the exchanger.

In accordance with the law of mass action [3], the thermodynamic constant of exchange equilibrium (3.12) has the form

$$K_B^A = \frac{a_{B^+} a_{AR}}{a_{A^+} a_{BR}}, \qquad (3.13)$$

where a_{A^+} and a_{B^+} represent the activities of the ions in solution [3], and a_{BR} and a_{AR} represent the activities of ions in the cationite.

The equilibrium constant K_B^A is a constant value for the investigated ion-exchange reaction (3.12) and it depends only on the temperature.

For cation exchange, written in the general form

$$z_B A^{+z_A} + z_A B^{+z_B} R_{z_B} \rightleftarrows z_A B^{+z_B} + z_B A^{+z_A} R_{z_A}, \qquad (3.14)$$

where z_A and z_B are the valences of ions A and B. The thermodynamic equilibrium constant takes on the form

$$K_B^A = \frac{a_B^{z_A} a_{AR}^{z_B}}{a_A^{z_B} a_{BR}^{z_A}}. \qquad (3.15)$$

If the numerical value of the equilibrium constant K_B^A is known, Eqs. (3.13) and (3.15) permit us to calculate the concentration of exchanging ions at equilibrium when the activity coefficients of the ions in the solution and in the ionite are known. The activity coefficients of the ions are calculated by different methods [3, 12–16].

Use of the law of mass action in ion exchange without consideration of the activity coefficients leads to the expression

$$K_c = \frac{C_B^{z_A} q_{AR}^{z_B}}{C_A^{z_B} q_{BR}^{z_A}}, \qquad (3.16)$$

where C_A and C_B are the concentrations of exchanging ions in solution, q_{AR} and q_{BR} are the concentrations in the resin phase, and K_c is the equilibrium coefficient [6].

The equilibrium coefficient is not a constant value. It depends on the concentration of the solution and the quantitative relations between ions A and B in the ionite. And only during exchange of univalent ions from dilute solutions is the equilibrium coefficient, as a first approximation, to be considered as possibly a constant value. Then the ratio of the average activity coefficients of the ions in the solution is approximately unity, and the resin phase may be assumed to be an ideal solid solution in which the activity is proportional to the ion concentration in the resin phase. Then the thermodynamic equilibrium constant K_B^A from Eq. (3.13)

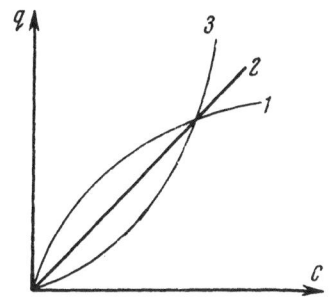

Fig. 9. Three types of ion-exchange isotherms: 1) convex; 2) linear; 3) concave

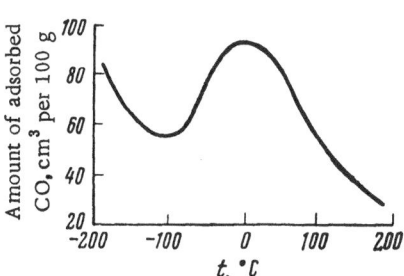

Fig. 10. Adsorption isobar of carbon monoxide by palladium.

[a value constant for the ion-exchange reaction (3.12)] will be near K_c from Eq. (3.16). Ion-exchange equilibrium is characterized also by an ion-exchange isotherm [6] that gives the relation of concentration of dissolved substance in the ionite to its concentration in the solution. The concentration is normally expressed in millimoles or milligram equivalents per gram of ionite and cubic centimeter of solution or in equivalent fractions. Typical ion-exchange isotherms are shown in Fig. 9. An analytical expression for an isotherm may be obtained from expression (3.16). For the exchange of equivalent ions here, it has been shown (see Fig. 9) that when $K_c = 1$ the isotherm is linear (2), when $K_c < 1$ it is convex (1), and when $K_c > 1$ it is concave (3). During exchange of ions with different valence, the shape of the isotherm is determined by the valencies of the exchanging ions.

The equilibrium constant of ion exchange depends weakly on temperature, and it diminishes as the temperature rises. This indicates that the termal effect of the ion-exchange reaction is small [6], no greater than 10 kcal/mole.

§ 23. The Concept of Chemisorption

Chemical adsorption, or chemisorption, in contrast to physical adsorption, discussed above, is accompanied by the formation of chemical bonds between the surface and the adsorbed molecules, i.e., by a redistribution of electrons between molecules of the adsorbate and the adsorbent. The nature of the compounds that form is determined primarily by the electron structure of the adsorbent [15].

The chemisorption of gases is observed on pure metallic surfaces, on coal, metal oxides, and other solids [15]. Chemisorption also takes place during heterogeneous catalysis [15]. Commonly, physical adsorption grades into chemical adsorption at higher temperatures. In Fig. 10 we have shown the isobar (P = const) of CO adsorption by palladium [5]. At low temperatures only physical adsorption occurs, decreasing as the temperature rises. Increase of adsorption, observed with further rise of temperature, is explained by the chemisorption of CO on palladium. At low temperatures chemisorption does not take place because it requires considerable activation energy [3, 15] (10-30 kcal/mole, as in chemical reactions, see Chap. 4). Diminution in amount of adsorbed substance at high temperatures is explained by rupture of chemical bounds between adsorbate and absorbent, i.e., desorption.

An appreciable thermal effect (liberation of more than 20,000 cal/mole) distinguishes chemisorption from physical adsorption. Elevation of temperature to a certain limit, also in contrast to physical adsorption, increases chemisorption (see Fig. 10). This is due to the fact that chemisorption is a chemical reaction, which requires notable activation energy. In some cases, however, it is impossible to draw a sharp boundary between the two types of adsorption.

Chemisorption isotherms have different shapes, depending on the system investigated. For various systems, chemisorption is described either by Eq. (3.5) of the Langmuir isotherm, or by the equation of Freundlich (3.1) and Temkin [16] (we have not examined the last).

§24. Ion-Exchange Equilibrium in Soils

Theoretical considerations and experimental studies on ion-exchange equilibrium in natural objects [8-10, 17-40] have been devoted chiefly to ion exchange in soils and minerals. Without dwelling on early works in ion exchange in minerals published in the past century (the works of Way, Lemberg, and others), let us consider ion exchange in soils. The principal aspects of ion exchange in soils are associated with the presence of the organic adsorbent complex in the soil [41]. However, the thermodynamics is ordinarily considered a process of ion exchange without regard to the bonding of the ions. Theoretical considerations, therefore, concerning ion-exchange equilibrium in soils, may be extended to ion exchange in rocks.

In considering ion exchange in soils to be a reversible reaction, Gapon proposed the following equation for exchange of univalent ions (3.13) [9]:

$$\frac{q_{AR}}{q_{BR}} = K_c \frac{C_A}{C_B}; \qquad K_c = \text{const.} \tag{3.17}$$

The exchange of ions having different valences [Eq. (3.14)] according to Gapon [9] conforms to the following relation:

$$\frac{q_{AB}}{q_{BR}} = K_c \frac{C_A^{1/z_A}}{C_B^{1/z_B}}. \tag{3.18}$$

This equation may be derived [27, 28] from the Poisson differential equation applied to the electrical double layer [5].

Nikol'skii proposed the following equation for the exchange of equivalent ions in soil [10, 17]:

$$\frac{q_{AR}^{1/z_A}}{q_{BR}^{1/z_B}} = K \frac{a_A^{1/z_A}}{a_B^{1/z_B}}. \tag{3.19}$$

Equation (3.19) may be obtained from Eq. (3.15) if we substitute concentration for activity of ions in the resin phase, raise both sides of the resulting equation to the power $\frac{1}{z_A \cdot z_B}$ and designate $(K_B^A)^{\frac{1}{z_A \cdot z_B}}$ by K. Use of $K = (K_B^A)^{\frac{1}{z_A \cdot z_B}}$ in place of the thermodynamic equilibrium constant K_B^A is convenient for a number of considerations. In Russian literature, therefore (in contrast to foreign literature), ion-exchange equilibrium is generally described by the thermodynamic equilibrium constant K.

In many works [9, 10, 17-22, 24-26], the equivalence and reversibility of exchange has been established for cations of the alkali and alkaline earth metals in soils. During exchange of ions having the same valence, the law of mass action in the form of Eq. (3.17) is fullfilled [19, 24-25]; i.e., the equilibrium coefficient K_c is constant for the given exchange. Numerical

TABLE 6. Numerical Values of the Equilibrium
Coefficient

Soil and rock	K_c	Reference
Chernozem soil	$K_{K, NH_4} = 2.617$	
	$K_{Ca, Mg} = 3.307$	Antipov-
Podsol soil	$K_{Ca, Mg} = 4.112$	Karataev
Terra rossa soil	$K_{Ca, Mg} = 4.055$	and others
Bentonite	$K_{K, NH_4} = 2.427$	[20, 21]
	$K_{Ca, Mg} = 2.159$	
Kaolin	$K_{K, NH_4} = 1.485$	
Clay soil	$K_{NH_4, K} = 1.37$	
Clay soil	$K_{NH_4, K} = 1.93$	Polyukov
Sandy loam soil	$K_{NH_4, K} = 1.90$	[24, 25]

values of K_c for the exchange $NH_4R + K^+ \rightleftarrows KR + NH_4^+$ in light loamy and sandy soils were found to be different (1.37 and 1.9 respectively [25]), which indicates that ion-exchange equilibrium depends on the properties of the soil. This latter is valid when we have to do with soils differing sharply in their physical and chemical properties.

Study of the exchange of cations of different valences [23] and also of heavy-metal cations (Cu^{2+}, Hg^{2+}) [21] in soils has shown that the law of mass action in the form of Eq. (3.17) is not fulfilled (the value of K_c in Eq. (3.17) stops being constant for the given exchange). As a rule this does not mean that cation exchange in soils is equivalent and reversible. Actually, if the ionic force of the electrolyte slowly changes during the course of the experiment, Eq. (3.17) is fulfilled [21]. This is explained by the fact that when the ionic force is constant, the activity values of the ions in solution may be replaced by concentrations. Eq.(3.19), becoming a stricter description of the law of mass action, then coincides with Eq. (3.17). An exception is the exchange of Ca in the soil for Hg^{2+} and Cu^{2+}, which does not conform to the law of mass action and is not reversible [21]. The Hg^{2+} and Cu^{2+} ions adsorbed by soil remain practically undisplaced by Ca^{2+} ions. In this case, mercury and copper are chemically adsorbed with the irreversible formation of chemical compounds on the surface of soil particles. The chemisorption of Ca^{2+} and Hg^{2+} is explained by their hydrolysis, as a result of which the copper and mercury are included among the anions that are adsorbed irreversibly. The phenomenon of irreversible adsorption of other anions (SO_4^{2-}, PO_4^{3-}) in soils has been discussed in a number of papers [37, 38, and others].

Numerical values of the equilibrium coefficient K_c, taken from a number of papers [19-22, 24, 25] for the exchange of different ions, are listed in Table 6.

It may be seen that the value of K_c differs little from unity. This indicates that the ion-exchange isotherms are nearly linear.

On the basis of their investigations, Antipov-Karataev and his coworkers [20-21] established the following series of ion adsorbability in chernozem soil:

$$Pb^{+2} > Ba^{+2} > Ca^{+2} > Mg^{+2}. \tag{3.20}$$

§25. Investigation of Adsorption and Ion Exchange in Rocks

Investigation of the interaction between natural minerals and solutions of electrolytes has shown that most minerals possess ion-exchange properties. As a rule, such minerals are crystalline silicates [6]. Among the minerals capable of cation exchange, zeolites must be given first mention: analcite $Na[Si_2AlO_6] \cdot H_2O$, chabazite $(Ca, Na)[Si_2AlO_6] \cdot 6H_2O$, harmotome $(K_2Ba)[Al_2Si_5O_{14}] \cdot 5H_2O$, heulandite $Ca[Si_3AlO_8]_2 \cdot 5H_2O$, and natrolite $Na_2[Si_3Al_2O_{10}] \cdot 2H_2O$. These minerals have a regular rigid crystal lattice in which some Si^{4+} ions are replaced by Al^{3+}. The deficiency of positive charge (because of the difference in charge between Al^{3+} and Si^{4+}) is compensated by ions of alkali or alkaline earth metals, which are not associated with any definite sites in the lattice. These cations are capable of exchanging with cations from the solution [6]. Zeolites have adsorption cavities, in which openings are small in comparison with the size of molecules. Adsorption by zeolites is therefore very sensitive to the size of the molecule being adsorbed. In this respect, zeolites are subdivided into separate types depending on the size of the entrance channels into the adsorption cavities.

Other cation exchangers are montmorillonite $Al_2[Si_4O_{10}(OH)_2] \cdot nH_2O$ and beidellite $Al_2[(OH)_2AlSi_3O_4OH] \cdot 4H_2O$, which in contrast to zeolites, have a loose structure [6]. They therefore swell considerably in water. Glauconites (ferro-aluminum silicates containing potassium) are also cation exchangers. They have a rigid lattice structure with small cavities. Cation exchange with these minerals therefore takes place chiefly on the crystal surface [6].

Some aluminosilicates are capable of exchanging both cations and anions. For example, montmorillonite, kaolinite, and feldspathoids such as sodalite and cancrinite [6] exchange hydroxyl ions for chlorine, sulfate, and phosphate ions. Anionites are apatite $[Ca_5(PO_4)_3]F$ and hydroxyapatite $[Ca_5(PO_4)_3]OH$. Some kinds of bituminous coal, most lignite, and anthracite also possess ion exchange properties [6].

Extensive investigations of the adsorption properties of rocks were made by Bykov and his coworkers [42-51] in connection with the search for inexpensive natural adsorbents for industrial use. Natural adsorbents are primarily highly dispersed rocks with specific surfaces reaching tens and hundreds of square meters. Bykov [42-43] distinguished the following principal types of natural adsorbents: 1) ashy tuffs and their weathering products, 2) agglomeratic tuffs of cinder cones and the deep weathering products of tuffs, 3) clays (montmorillonitic, bentonitic, kaolinitic), the products of reworking weathered material from igneous rocks, and 4) diatomite. The specific surface of kaolinitic clays ranges from 17 to 65 m^2/g; bentonic clays, from 40 to 96 m^2/g; and ashy tuffs, from 20 to 95 m^2/g. These rocks are good adsorbents and are being used in industry. The high adsorption capacity of decomposed ashy and agglomeratic tuffs is due to the formation of clay minerals in their pores—halloysite and montmorillonite — as a result of weathering. As investigations have shown, the adsorption capacity of tuffs decreases with increase in depth of occurrence, which follows from the decline in weathering of tuffs with depth.

Materova [52] made detailed investigations of the exchange of Na^+, Li^+, Ag^+, and Ba^{2+} in glauconite that has been converted to the H^+-form. The exchange of cations takes place at the expense of the negatively charged glauconite because of dissociation of surface molecules. A double layer is formed on the surface, the cations of the layer exchanging with cations from the solution. The effect of the nature of the ion on its adsorption is determined by the lyotropic series. The exchange capacity of glauconite is rather appreciable (on the order of 15-24 mg equiv per 100 g). Anion exchange does not occur with glauconite.

Materova [52] demonstrated the equivalence and reversibility of cation exchange in glauconite. She showed that the cation exchange conforms to the law of mass action in the form of (3.19). A study of exchange at different pH values of the solution showed that the amount of adsorbed cations increases linearly with increase of pH.

Ermolenko and Shirinskaya [53] studied the exchange adsorption of ions of alkali and alkaline earth metals (Na^+, K^+, Ca^{2+}, and Ba^{2+}) on different clays, taken in the calcium and barium forms. It was found that the exchange capacity of Ca^{2+}-clays is 68 mg equiv per

100 g of clay; the exchange capacity of Ba^{2+}-clays is 23.4 mg equiv per 100 g. In studying the exchange of Na^+ and K^+ on Ca^{2+} clays, it was ascertained that K^+ is adsorbed better than Na^+. In all cases, Ca^{2+} was desorbed from clay in greater quantities than Ba^{2+}, which points to selective adsorption by clay of the alkaline earth metals from mixtures ($Ca^{2+} > Ba^{2+}$). The authors found that the exchange of univalent and multivalent cations on clays is equivalent and reversible.

Investigation of the adsorption of alkali and alkaline earth metals on clays of different composition (montmorillonitic, kaolinitic) was carried out by Antipov-Karataev and his coworkers [19-22]. Qualitatively, the results are similar to those obtained in the study of exchange in soils. The series of cation adsorbability on montmorillonitic clay is the following:

$$Pb^{+2} > Ca^{+2} > Ba^{+2} > Mg^{+2}. \tag{3.21}$$

A number of authors [54] have studied the adsorption of $CuSO_4$ on the following minerals: pyrite, chalcopyrite, galena, smithsonite, and calamine as a function of the pH of the solution. The studies were made in connection with the fact that $CuSO_4$ is adsorbed by these minerals in the flotation process. It was found that at low concentrations of $CuSO_3$ (< 3.2 mg/liter) the adsorption isotherm is linear. The linearity disappears with increasing concentrations. The dependence of the adsorption on pH of the solution is complex and is characterized by the presence of a maximum. In pyrite, for example, the adsorption maximum is observed at a pH of 9.5, in chalcopyrite at 7.0.

Kokotov and others [55] studied the adsorption of Sr^{90} and Cs^{137} (two long-lived products of the nuclear fission of uranium) by different soils and clays as a function of the presence of other ions and of the pH of the solution. The purpose of the investigation was to show the amount of adsorption of these isotopes by soils and in what way they might be removed from root-producing layer. It was found that Sr^{90} is adsorbed reversibly. The dependence of the distribution coefficient on pH of the solution is characterized by one or two maximums. When Ca^{2+} ions are added to a solution containing Sr^{90}, the number of adsorbed Sr^{90} begins to decline at a concentration of $Ca^{2+} > 0.01 N$. This is due to the fact that at low concentrations of the Ca^{2+} ion there is no competition with adsorbed strontium. When the amount of Ca^{2+} in solution exceeds the adsorption capacity of the soil, then Ca^{2+} is adsorbed chiefly and it displaces Sr^{90} from the soil. Consequently, Sr^{90} may be removed from the root-producing layer by flushing it with a rather concentrated solution of salt (such as a solution of $CaCl_2$ with a concentration of $0.01 N$ or greater). A study of the adsorption of Cs^{137} by a turfy podsol soil, southern chernozem, and kaolin has shown that soils possess great adsorption capacity [the distribution coefficient — the ratio of number of adsorbed ions to the number of ions in solution — $K \approx (1.0-1.6) \cdot 10^7$]. In this process, cesium is probably adsorbed irreversibly. The amount of adsorbed Cs^{137} decreases in the presence of NH_4^+, K^+, and other ions. However, because of large values of K, the adsorbed Cs^{137} is more difficult to remove from the soil by flushing with salt solutions than Sr^{90}.

S. K. De [56] studied the adsorption of phosphates $(NH_4)_3PO_4$, $(NH_4)_2HPO_4$, $(NH_4)H_2PO_4$, and H_3PO_4 on montmorillonite taken in the NH_4^+-, Na^+-, K^+-, and Li^+- forms. It was found that the different forms of montmorillonite may be arranged in a series according to decreasing adsorption capacity:

$$NH_4 > Na > K > Li. \tag{3.22}$$

The adsorption decreased also with increase of pH of the phosphate solution, and in all cases it bore an exchange character, since different amounts of Al^{3+}, Ca^{2+}, and Mg^{2+} were displaced from the mineral during preparation of the cation forms.

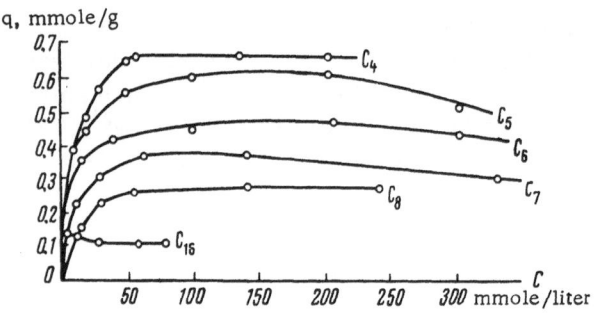

Fig. 11. Adsorption isotherms of normal alcohols on decomposed ashy tuff from a solution in hexane. (The subscript of the symbol C indicates the number of carbon atoms in the alcohol molecule.)

Ion exchange on minerals was studied by the authors whose works were discussed above. Molecular adsorption of organic substances on rocks was studied by Bykov and others in order to understand the adsorption properties of rocks. Figure 11 shows the adsorption isotherms of a homologous series of alcohols from hexane on decomposed ashy tuff. With increase in molecular size of the adsorbed substance, adsorption declines because of the decrease in number of pores accessible for adsorption.

Similar results were obtained for the adsorption of fatty acids from hexane and normal alcohols from carbon tetrachloride on different rocks (tuffs, clays, diatomites).

§ 26. Adsorption of Gases on Rocks

Experimental studies of the adsorption of gases on rocks have been undertaken chiefly in connection with the search for inexpensive natural adsorbent for industrial use [46-51] and in order to interpret the results of "gas surveys" [57].

Vasserberg [57] investigated the adsorption of saturated hydrocarbons (methane, ethane, propane, butane, and pentane) on different rocks from the surface (depths of 2-3 m) and from various drill-hole depths. The rocks were chiefly clays, sandstones, and limestones. Adsorption for all investigated vapors proved to be small. The amount of adsorbed methane was between 0.04 and 0.1 ml of gas per gram of adsorbent at pressure of 760 mm Hg. With increase in molecular weight of adsorbed gas, the adsorption increased, especially at the change from butane to pentane.

Vasserberg explained this by saying that the boiling point of pentane is low (36°C). At the high temperatures of the experiment, capillary condensation of the adsorbed substance began to be effective. The author was not able to establish a single-valued relationship between the adsorption capacity of a rock and its petrographic characteristics. It might be said, merely as a first approximation, that the adsorption capacity of clays is greater than that of limestones. The study of adsorption of methane on rocks at various temperatures showed that adsorption declines with increase in temperature.

The conclusions expressed in the indicated paper [57] may be summarized as follows: 1) the lack of a single-valued correspondence between the adsorption capacity of rocks and their petrographic character makes it desirable to study the adsorption capacity of the principal rock-forming minerals, since it has not been possible to compute the adsorption capacity of complex rocks from the available data; 2) the current of hydrocarbon gases diffusing from a deposit to the surface is impoverished in the higher hydrocarbons, which are capable of capillary condensation; the "heavy fraction" of gases collected at the surface contains only small quantities of hydrocarbons above propane; 3) since the adsorption of light hydrocarbons is small, it has been possible for equilibrium between gas and rock to be reached within geologic time, after which the amount of diffusing hydrocarbons remains constant.

The above conclusions are qualitative. The quantitative effect of adsorption on diffusion will be examined in Chapter 6.

Bykov and his coworkers [46-51] investigated the adsorption of vapors of organic substances on different natural adsorbents in order to examine structural types of adsorbents. The

adsorption isotherms (circles) and desorption isotherms (dots) of benzene vapor on different
tuffs and diatomites, the weathering products of agglomeratic tuffs of Early Quaternary volca-
noes, are illustrated in Fig. 12. All the isotherms are S-shaped, which is explained by capillary condensations of benzene vapor. It was shown that adsorption is greater in the decomposed
ashy tuffs. Analysis of the isotherms by the BET (Brunauer-Emmett-Teller) method shows
that the specific surface of ashy tuffs is on the order of 25-35 m^2/g, but of decomposed
ashy tuffs, 38-40 m^2/g. Decomposed agglomeratic tuffs have higher adsorption properties, a
specific surface of 70-90 m^2/g. In their adsorption capacity, diatomites are similar to lipar-
itic ashy tuffs and their weathering products.

A number of authors [47-48] have investigated the adsorption of water and heptane vapors
on samples of decomposed ashy tuff. The adsorption isotherms are S-shaped. Figure 13 shows
that the adsorption isotherms (circles) and desorption isotherms (dots) do not coincide throughout the curve (irreversible hysteresis). This is explained, in all probability, by chemical adsorption of water vapor. Investigation of heptane adsorption on rocks of different moisture contents
has shown [48] that adsorption diminishes with increase of moisture. Within the range of 5-20%
moisture content, the decrease of adsorption follows a linear law. This is due to the adsorption
of water, leading to a decrease in surface of the adsorbent accessible for adsorption of heptane
molecules.

Bykov and others [49] studied the adsorption of water vapor and the homologous series
of alcohols (methyl, ethyl, n-propyl). Since the molecules of water and alcohol have different
diameters, these data on adsorption may be used to evaluate the structure of the adsorbent
(Dubinin's method of "molecular gages"). With increase in size of the molecules of the adsor-

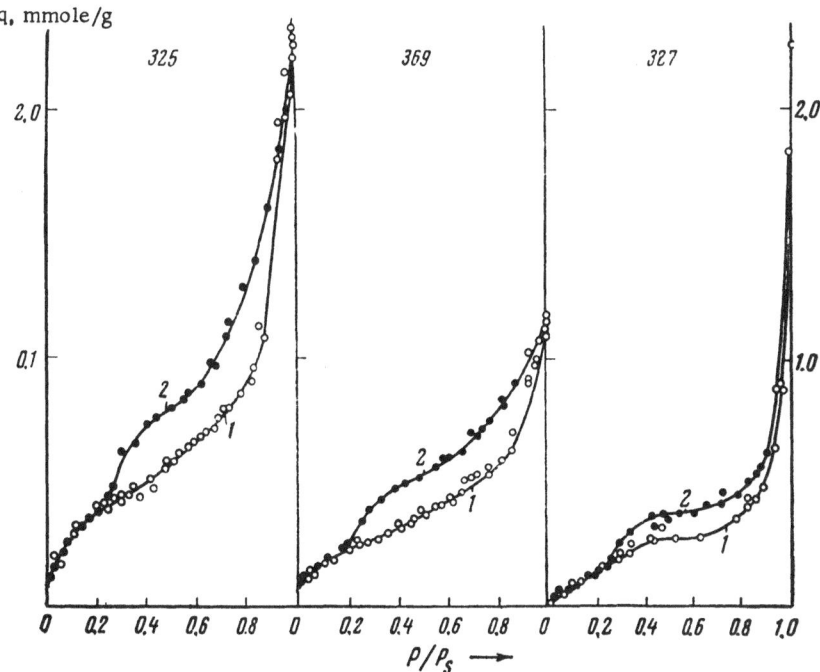

Fig. 12. Adsorption isotherms (1) and desorption isotherms (2) of
benzene on samples of natural adsorbents: 325) agglomeratic tuff;
369) kaolinite; 327) diatomite.

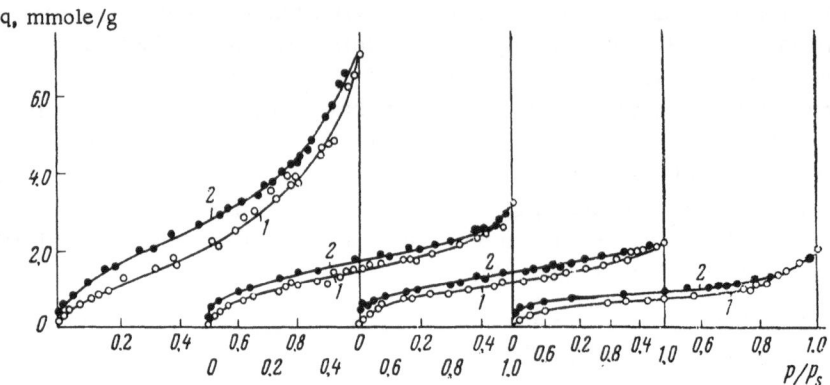

Fig. 13. Isotherms of adsorption (1) and desorption (2) of water vapor and alcohol vapor on ashy tuff.

bate, adsorption declines, since the number of pores accessible to the molecules diminishes.
Consequently, pores of different sizes are present in rocks. The lack of coincidence of the
adsorption and desorption isotherms, any of them, as obtained by experiment, is probably due
to chemical adsorption through interaction between alcohol vapor and the rock surface.

In one paper [5], the adsorption of benzene on montmorillonite and kaolinite at different
temperatures (20, 70, 120, 170 and 220°C) was discussed. The adsorption isotherm of benzene
on montmorillonite changes shape with rise in temperature, changing from convex to concave;
the maximum value of adsorption declines. The irreversibility of adsorption is explained [5]
by the swelling of montmorillonite in benzene; as the temperature rises this adsorption disappears. These phenomena may be due to changes in the mechanism of adsorption with increase
in temperature as a result of the emission of structural water of the adsorbent.

The main difference between the adsorption isotherms of benzene vapor on kaolinite from
the isotherms on montmorillonite is that the adsorption on kaolinite is reversible at any temperature. With rise in temperature, the maximum value of adsorption increases because of removal of adsorbed water.

From the papers discussed above [42-51], it follows that different rocks differ in their
adsorption properties, and the difference may be large. Bykov, in particular, introduced a
classification and rating of minerals according to their adsorption properties [51]. The state-
ment of Vasserberg [57] concerning the lack of any connection between adsorption capacity of
rocks and their petrographic character is not precise, therefore.

LITERATURE CITED

1. Brunauer, S., The adsorption of Gases and Vapors, Vol. 1, Princeton University Press,
    Princeton (1943).
2. De Boer, J. H., The Dynamical Character of Adsorption, The Clarendon Press, Oxford
    (1953).
3. Gerasimov, Ya. I., and others, A Course in Physical Chemistry [in Russian], Vols. 1 and 2,
    Izd. Khimiya (1963, 1966).
4. Dubinin, M. M., and Radushkevich, L. V., Dokl. Akad. Nauk SSSR, Vol. 55, No. 4, p. 331
    (1947).
5. Boyutskii, S. S., A Course in Colloidal Chemistry [in Russian], Izd. Khimiya (1964).
6. Helfferich, F., Ion Exchange, McGraw-Hill, New York, (1962).
7. Saldadze, K. M., Pashkov, A. B., and Titov, V. S., High-molecular Ion-Exchange Com-
    pounds [in Russian], Izd. Khimiya (1960).

8. Lans, R., Zentralblatt Geolog. Paläont., p. 699 (1913).

9. Gapon, E. N., Zh. Org. Khim. No. 3, p. 144 (1933); No. 7, p. 1438 (1937).

10. Nikol'skii, B. P., Pochvovedeniye, No. 2 (1934).

11. Kielland, J., J. Soc. Chem. Ind., Vol. 54, p. 2321, London (1935).

12. Boyd, G. E., Schubert, J., and Adamson, A. W., in: The Chromatographic Method of Separating Ions [Russian translation], Izd. Inostr. Lit. (1949).

13. Ekedahl, E., Högfeldt, E., and Sillén, L. G., Acta. Chem. Scan., Vol. 4, p. 556 (1950).

14. Argensinger, W. J., Jr., Davidson, A. W., and Bonner, O. D., Trans. Canad. Acad. Sci., Vol. 53, p. 404 (1950).

15. Trepnel, B., Chemisorption [Russian translation], Izd. Inostr. Lit. (1958).

16. Frumkin, A., and Shlygin, A., Acta Physicokhim. USSR, Vol. 3, p. 791 (1935).

17. Nikol'skii, B. P., and Paramonova, V. I., Uspekhi Khimii, Vol. 8, p. 1535 (1939).

18. Davydov, A. T., and Skoblionok, R. F., Tr. Inst. Khimii, KhGU, Vol. 10, p. 107 (1952).

19. Antipova-Karataeva, T. F., and Antipov-Karataev, N. N., Pochvovedenie, No. 2, p. 52 (1940).

20. Antipov-Karataev, I. N., and Kader, G. M., Kolloid. Zh., Vol. 9, p. 81 (1947); Vol. 9, p. 161 (1947).

21. Antipov-Karataev, I. N., Kader, G. M., and Fillipova, V. N., Kolloid. Zh., Vol. 9, p. 315 (1947); Vol. 10, p. 73 (1948).

22. Antipov-Karataev, I. N., Paskvik-Khlopina, M. A., Merkulova, M. S., and Grebenshchikova, V. I., Kolloid. Zh., Vol. 10, p. 401 (1948).

23. Chernov, V. A., Tr. Pochvennogo Inst. im. Dokuchaeva, No. 9, p. 13 (1939); No. 10, p. 106 (1940).

24. Polyakov, Yu. A., Kolloid. Zh., Vol. 9, p. 439 (1947).

25. Polyakov, Yu. A., Tr. Pochvennogo Inst. im. Dokuchaeva, Vol. 51, p. 158 (1956).

26. Krishnamoorthy, C., and Overstreet, R., Soil Sci., Vol. 69, p. 41 (1950).

27. Erikson, E., Soil Sci., Vol. 72, p. 103 (1952).

28. Bolt, G., Soil Sci., Vol. 79, p. 267 (1955).

29. Mortland, M. M., Soil Sci., Vol. 80, No. 1 (1955).

30. Goring, C., and Martin, R., Soil Sci., Vol. 88, p. 336 (1958).

31. Huffaker, R., and Wallace, A., Soil Sci. Vol. 87, p. 130 (1959).

32. Bower, C., Soil Sci., Vol. 88, p. 32 (1959).

33. Tendille, Chande Eynard, C. R. Acad. Sci., Vol. 248, No. 18, p. 2638 (1959).

34. Timofeeva, N., and Titlyanova, A., Izv. Akad. Nauk SSSR, Ser. Biol., p. 111 (1959).

35. Coleman, N., Thorup, I., and Jackson, W., Soil Sci., Vol. 90, p. 1 (1960).

36. Brown, J., and Jones, W., Soil Sci., Vol. 94, p. 171 (1962).

37. Fang, S., Nance, R., Tsun Tien Chao, and Harward, M., Soil Sci., Vol. 94, p. 14 (1962).

38. Tsun Tien Chao, Harward, M., and Fang, S., Soil Sci., Vol. 94, p. 276 (1962).

39. Tiller, H., Hodgson, Y., and Peech, M., Soil Sci., Vol. 95, p. 392 (1963).

40. Ramachandran, V., Kacker, H., and Rao, H., Soil Sci., Vol. 95, p. 414 (1963).

41. Gedroits, K. K., Study of The Adsorption Capacity of Soil [in Russian] (1934).

42. Bykov, V. T., Dokl. Akad. Nauk SSSR, Vol. 79, p. 621 (1951).

43. Bykov, V. T., and Smirnova, L. V., Tr. DV Filiala AN, Ser. Khimich., No. 3 (1958).

44. Bykov, V. T., and Yakovleva, M. V., Tr. DV Filiala AN, Ser. Khimich., No. 4, p. 68 (1960).

45. Bykov, V. T., Sakhno, V. G., and Ustinovskii, Yu. B., Tr. DV Filiala AN, Ser. Khimich., No. 4, p. 5 (1960).

46. Gerasimova, V. G., and Bykov, V. T., Tr. DV Filiala AN, Ser. Khimich., No. 4, p. 41 (1960).

47. Gerasimova, V. G., Tr. DV Filiala AN, Ser. Khimich., No. 3, p. 94 (1958).

48. Gerasimova, V. G., and Bykov, V. T., Tr. DV Filiala AN, Ser. Khimich., No. 3, p. 109 (1958).

49.   Bykov, V. T., and Smirnova, L. V., Tr. DV Filiala AN, Ser. Khimich., No. 4, p. 55 (1960).

50.   Bykov, V. T., Gor'kovskaya, V. T., and Frolov, B. A., Tr. DV Filiala AN, Ser. Khimich. No. 3, p. 52 (1958).

51.   Bykov, V. T., Tr. DV Filiala AN, Ser. Khimich., No. 4, p. 34 (1960).

52.   Materova, E. A., Uch. Zap. LGU., No. 7, p. 15 (1945).

53.   Ermolenko, N. F., and Shirinskaya, L. N., Izv. Vyssh. Uch. Zaved., Khimiya i Khim. Tekhnolog., Vol. 5, p. 468 (1962).

54.   Mitrofanova, S. N., and Kushnikova, V. G., Sb. Tr. Gos. NII Tsvetn. Metal., No. 19 (1962).

55.   Kokotov, Yu. A., Popova, R. F., and Urbanyuk, A. P., Radiokhimiya, No. 2, p. 200 (1961).

56.   De, S. K., and Ind, I., Soil Sci. Soc., Vol. 9, p. 169 (1961).

57.   Vasserberg, V. É., in: Geochemical Methods of Prospecting for Oil [in Russian], p. 117 (1950).

CHAPTER 4

THE KINETICS OF ADSORPTION, ION EXCHANGE, AND CHEMICAL REACTIONS OF SOLUTIONS AND GASES WITH ROCKS

As pointed out in Chap. 1, the geochemical migration of substances is described by a system of equations of material balance and the kinetics of interaction processes (adsorption, ion exchange, and chemical reactions) of the substances with rocks under certain initial and boundary conditions. Consequently, the interaction rate of solutions and gases with the rocks determines the laws of geochemical migration (Chaps. 6 and 7). This makes it important to consider the kinetics of adsorption, ion exchange, and chemical reactions of solutions and gases with rocks.

§ 27. The Basic Aspects of Formal Kinetics

Let us examine the general form of the chemical reaction

$$\nu_1 A_1 + \nu_2 A_2 + \ldots + \nu_n A_n \rightarrow \nu_1' A_1' + \nu_2' A_2' + \ldots + \nu_n' A_n', \tag{4.1}$$

where A_1 and A_1' are the initial and resulting substances, and ν_1 and ν_2 are the stoichiometric coefficients.

By rate of chemical reaction we mean the number of molecules or gram molecules of a given substance reacting per unit time in unit volume [1]. To measure the reaction rate it is sufficient to determine the change with time of one of the participating components of the reaction, since changes in the amounts of remaining substances may be found on the basis of a stoichiometric equation of the reaction (4.1).

In accordance with this determination, the rate of chemical reaction v is

$$v = - \frac{dn_{A_1}}{V\,dt}, \tag{4.2}$$

where n_{A_1} is the number of gram molecules of substance A_1 at time t, and V is the volume of the system.

If the volume of the system does not change (reactions in solutions), then we shall have in place of Eq. (4.2)

$$v = \frac{dC_{A_1}}{dt}, \tag{4.3}$$

where C_{A_1} is the concentration of the substance A_1.

According to the basic postulate of chemical kinetics [1], the reaction rate (4.1) is

$$v = K C_{A_1}^{v_1} C_{A_2}^{v_2} \ldots C_{A_n}^{v_n},$$ (4.4)

where C_{A_1} (i = 1, 2, ..., n) represents the concentrations of the reacting substances, and K is the constant of the reaction rate (4.1).

The physical meaning of K follows from Eq. (4.4); if we assume

$$C_{A_1} = C_{A_2} = \ldots C_{A_n} = 1,$$ (4.5)

then

$$v = K.$$

Thus, the constant of the chemical-reaction rate is equal to the rate of this reaction under conditions such that the concentrations of the reacting substances are constant and equal to unity.

Chemical reactions may be divided into reversible and irreversible. Reversible reactions are those that simultaneously and independently take place in both forward and reverse directions. Irreversible reactions are those that take place in only one direction. Reversible reactions tend toward establishment of chemical equilibrium, at which condition the rates of forward and reverse reactions become adjusted. Irreversible reactions proceed to the end, i.e., to complete loss of the original substances. All chemical reactions are reversible. However, under certain conditions, the reactions may proceed in only one direction, i.e., be irreversible.

In order for the elemental act of chemical reaction to take place, reacting particles (molecules, atoms, ions) must collide with each other. We distinguish monomolecular, bimolecular, and polymolecular reactions according to the number of reacting particles. Since the probability of simultaneous collision of many molecules is small, polymolecular reactions are not highly probable. In practice, we have to do only with reactions of the first, second, and, rarely, third orders.

We shall find, on the basis of Eq. (4.4), rate equations for some reactions. For simplicity let us assume that the volume of the system does not change with time.

1. Irreversible Reaction of the First Order. The general form of this reaction may be written

$$A \rightarrow v_1' A_1' + v_2' A_2' + \ldots .$$ (4.6)

In accordance with Eqs. (4.3) and (4.4), we obtain the following differential equations of the reaction rate (4.6):

$$-\frac{dC}{dt} = KC,$$ (4.7)

where C is the concentration of substance A at time t.

Let the concentration of substance A be C_0 at zero time, t = 0. Integrating Eq. (4.7) for this initial condition, we obtain

$$C = C_0 e^{-Kt}.$$ (4.8)

The integral equation (4.8) of the reaction rate (4.6) describes the change in concentration of the decomposing substance A with time.

2. Irreversible Reaction of the Second Order. The general form of this reaction may be written

$$A_1 + A_2 \rightarrow v_1' A_1' + v_2' A_2' + \ldots . \tag{4.9}$$

In accordance with Eq. (4.4), the differential equation of the reaction rate (4.9) has the form

$$-\frac{dC_1}{dt} = -\frac{dC_2}{dt} = K C_1 C_2, \tag{4.10}$$

where C_1 and C_2 are the concentrations of substances A_1 and A_2 at time t.

Integrating Eq. (4.10) for the case when the concentrations of reacting substances are equal ($C_1 = C_2 = C_3$), we obtain

$$C = \frac{1}{C_0^{-1} + Kt}, \tag{4.11}$$

where C_0 represents the initial concentrations of A_1 and A_2.

3. Reversible Reaction of the First Order. This is a reaction of the type

$$A \xrightarrow[K_2]{K_1} A'. \tag{4.12}$$

The reaction rate of (4.12) is equal to the difference in rates of the forward and reverse reactions, each of which is monomolecular. In accordance with Eq. (4.7), we shall then have

$$-\frac{dC_A}{dt} = K_1 C_A - K_2 C_{A'}, \tag{4.13}$$

where K_1 and K_2 are constants of the rates of forward and reverse reactions, respectively.

Integration of Eq. (4.13) is not very difficult [1].

4. Reversible Reaction of the Second Order. The general form of this equation may be written

$$A_1 + A_2 \xrightarrow[K_2]{K_1} A_1' + A_2'. \tag{4.14}$$

The differential equation of rate has the form

$$-\frac{dC_{A_1}}{dt} = K_1 C_{A_1} C_{A_2} - K_2 C_{A_1'} C_{A_2'}. \tag{4.15}$$

If the rate of reversible reaction conforms to the monomolecular law, i.e.,

$$A_1 + A_2 \xrightarrow[K_2]{K_1} A', \tag{4.16}$$

then in place of Eq. (4.15), we have

$$-\frac{dC_{A_1}}{dt} = K_1 C_{A_1} C_{A_2} - K_2 C_{A'}. \tag{4.17}$$

For more complex reactions, the differential equations of rate are written in similar manner.

§ 28. Heterogeneous Processes

The above discussion refers chiefly to homogeneous chemical reactions, taking place in a single phase. In heterogeneous processes, the reacting substances are in different phases, so that the reaction occurs at an interface between phases. By virtue of this, complicating factors arise, associated with the transport of substances to the zone of reaction. During migration of solutions in the earth's crust, heterogeneous chemical reactions between migrating substances and the enclosing rocks represent the dominant processes. For geology, therefore, the kinetics for the rules of heterogeneous processes are of special interest.

Any heterogeneous reaction involves several stages. In particular, when a liquid or gaseous solution reacts with a solid on its surface, the process involves: 1) approach of the substance to the surface, 2) the act of chemical reaction, and 3) withdrawal of the substance, formed by reaction, in the bulk solution. The processes of transporting substances result from differences in concentrations in the bulk solution and on the reaction surface: molecular (ionic) diffusion or, when the reacting substances are agitated or flow, convective diffusion [2].

The overall rate of the heterogeneous process is determined by the rates of the individual stages. However, if the rate of one of the stages is much lower than the rate of the others, the rate of the process is determined by the rate of the slowest stage. If the rate of chemical reaction is lower than the diffusion rate, the process is said to lie in the kinetic region. The rate of the process in this case is described by an equation of the rate of reaction taking place at the phase interface, and this is found from the basic postulate of chemical kinetics (4.4). On the other hand, if the diffusion rate is lower than the rate of chemical reaction, the process takes place in the diffusion region. When the diffusion rate and the rate of chemical reaction are similar, the process lies in transitional (mixed) regions.

A particular reaction, depending on the conditions under which it takes place (migration rate, temperature, and so forth), may lie in different kinetic regions. For example, it was found that the diffusion rate diminishes by a factor of 1.2 when the temperature is lowered 10°, but the reaction rate is reduced by a factor of 3-4. Consequently, if the process takes place in the diffusion region, when the temperature is lowered, the process shifts to the kinetic region. Several methods exist for determining the region in which heterogeneous processes take place. We shall dwell on this later.

Let us examine the classic view concerning diffusion kinetics of the reaction between a liquid solution and a solid. The theory of this problem on the basis of the solution of solids in liquids was worked out first by Nernst. According to Nernst's theory, a saturated solution of the substance arises in the layer of liquid at the surface of the body. Solution takes place by diffusion of substance from this layer into the bulk solution. If we use δ to denote the distance at which the concentration is changed from the concentration of saturated solution C_s to the exchange concentration C_0, then, in accordance with Fick's law, (2.1), we may write the following expression for the amount of substance dissolving per unit time:

$$Q = D \frac{C_s - C_0}{\delta} S, \qquad (4.18)$$

where S is the surface area on which solution is taking place, and δ is the thickness of the so-called Nernst diffusion layer.

If solution takes place with agitation or flow of the solvent, then, according to Nernst, a fixed layer of liquid with a thickness of δ becomes attached to the surface of the solid, diffusion of dissolved substance taking place in this layer. The concentration of the substance is constant at the boundaries of this layer because of agitation. It has been found experimentally that $\delta \approx 10^{-2} - 10^{-4}$ cm [2], and it decreases with increase in flow rate of the solvent according to the law

$$\delta \approx \frac{1}{u^n}, \quad n \approx 0.5 - 1. \tag{4.19}$$

Nernst's theory is a qualitative theory of diffusion kinetics of heterogeneous processes, since it furnishes no theoretical expression for the thickness of the diffusion layer δ . Consequently, no absolute value of the diffusion current Q can be calculated theoretically.

Levich [2] has examined several questions on the quantitative theory of diffusion kinetics of heterogeneous chemical reactions on the basis of solution of the equation of convective diffusion with appropriate initial and boundary conditions. This theory has been applied by one of the authors of this book jointly with G. M. Panchenkov to the phenomena of adsorption and ion exchange [3-4]. In Chap. 5 we shall examine the quantitative theory of kinetics of heterogeneous processes on the basis of adsorption of liquids or gas by solid adsorbents from a current, and we shall show how this differs from Nernst's theory.

§ 29. Kinetic and Diffusion Regions of Heterogeneous Chemical Reactions

Let us introduce an equation for the reaction rate between solution and solid when the process takes place in the intermediate or transitional region [1]. The diffusion rate v_1, in accordance with the concept of the Nernst diffusion layer, may be written

$$v_1 = \gamma [C - C_1], \tag{4.20}$$

where C is the concentration of substance in the bulk solution, C_1 the concentration on the surface of the body, and γ the constant of diffusion rate, depending on the diffusion coefficient and the thickness of the diffusion layer.

The rate of chemical reaction v depends on the concentration C_1 on the surface in accordance with Eq. (4.4). For the reaction of the first order

$$v_2 = KC_1. \tag{4.21}$$

When the course of the process is established, the diffusion and reaction rates become equal to each other:

$$KC_1 = \gamma (C - C_1). \tag{4.22}$$

Substituting C_1, found from Eq. (4.22), in Eq. (4.21), we obtain the reaction rate v in the steady state

$$v = v_1 = v_2 = K_1 C, \tag{4.23}$$

where

$$\frac{1}{K_1} = \frac{1}{K} + \frac{1}{\gamma}. \tag{4.24}$$

From Eqs. (4.23) and (4.24) it follows that in this case the overall rate of the heterogeneous process corresponds to the first order for the concentration C of reacting substance, and the reciprocal value of the rate constant of the process is equal to the sum of the reciprocal values of the rate constants of the separate stages. When the order of the reaction on the surface is higher than the first, we obtain more complicated relations [1].

§ 30. The Kinetics of Adsorption and Ion Exchange

Adsorption and ion exchange are typical heterogeneous processes. The general transfer of substance between a grain of the adsorbent and the solution may be divided into the following stages: 1) diffusion of the adsorbate from the bulk solution to the grain of the adsorbent, 2) diffusion into the grain, 3) adsorption or ion exchange, 4) diffusion of the substance formed by ion exchange to the surface of the grain, and 5) diffusion of the substance into the bulk solution. Numerous experimental investigations of adsorption (ion exchange) have shown that the rates of the processes are generally determined by diffusion. Adsorption (ion exchange) may therefore take place in three limiting regions: 1) external-diffusion kinetics, when the rate of adsorption (ion exchange) is limited by the diffusion of substance to the surface of the adsorbent, 2) internal-diffusion kinetics, when the adsorption rate is limited by diffusion into the grain, and 3) mixed diffusion kinetics, when the rates of substance transfer are equal to each other.*

Depending on the conditions under which the experiment is perfomed, adsorption and ion exchange may take place in any diffusion region. Experimental studies of ion exchange on synthetic exchangers [5] have established that at low concentrations ($C \leq 0.003N$) of the exchanging ion, the process takes place in the external-diffusion region. At greater concentrations ($C \geq 0.5N$ it takes place in the internal-diffusion region. In the intermediate range of concentrations, ion exchange takes place in the mixed diffusion region. These simple relationships are not observed for adsorption [6]. The region in which the process takes place is determined to a considerable extent by the properties of the adsorbent itself. Large-pore adsorbents, for example, generally provide adsorption in the external-diffusion region, but finely porous adsorbents exhibit internal-diffusion adsorption.

In deriving kinetic equations describing adsorption in the external-diffusion region, it is assumed, in keeping with experimental data, that adsorption equilibrium obtains on the surface of the adsorbent. As a consequence, the concentration of the solution $\overline{C}$ immediately adjacent to the suface is related to the concentration q of adsorbed substance by the equation of the adsorption isotherm (ion-exchange isotherm) $\overline{C} = f(q)$. Then, in accordance with the qualitative theory of Nernst, we may write:

$$\frac{dq}{dt} = \gamma \, [C_0 - f(q)], \qquad (4.25)$$

where γ is the rate constant of external diffusion, depending on D, δ, and the geometric parameters of the adsorbent.

A theoretical expression for γ will be found in the following chapter, when adsorption from solution flow is considered.

The transfer of adsorbate substance within the adsorbent takes place through pores, and, as a consequence of the different mechanisms of transferring substances, it is a complex pheno-

---

*Translator's note: In granular porous rocks, with some diffusion through the pores, some through the grains themselves, the authors' term external diffusion may be sensibly called intergranular diffusion, and internal diffusion may be termed intragranular. However, the latter terms cannot be universally substituted for external and internal, because, in rocks with negligible pore space, external diffusion may refer to movement through major fractures, internal diffusion to movement through microfractures. Nor can intergranular and intragranular be properly applied to masses of organic material.

menon. As we pointed out in Chap. 2, with any type of transfer of adsorbate substance in pores, including transfer in rocks and soils, the rate of the process may be formally expressed by the diffusion equation with some effective value for the diffusion coefficient D_{ef}. A clear form of the equation of internal-diffusion kinetics of adsorption will be found in § 31 for adsorption in grains of different forms (cylinders, spheres).

At least two ions, A and B, participate in ion exchange. To derive kinetic equations, it is therefore necessary to consider two currents: the flow of ions A toward the grains of the rock and the flow of ions B from the grains. By virtue of the requirements of electroneutrality of ion exchange [5], these currents are equal at any instant of time. As shown in some papers [4, 7], kinetic equations of adsorption [such as (4.25)] may be applied to ion exchange. However, the coefficients of molecular diffusion D and D_{ef} must be substituted for the coefficients of interdiffusion of ions A and B, which are approximately equal [5] to

$$\frac{1}{D} = \frac{1}{D_A} + \frac{1}{D_B}, \tag{4.26}$$

where D_A and D_B are individual diffusion coefficients of ions A and B.

Thus, ion exchange in the diffusion regions may be considered similar to reversible molecular adsorption, so that in theory the kinetic equations will be derived henceforth for adsorption and ion exchange simultaneously.

§ 31. Equations of Internal-Diffusion Kinetics of Adsorption and Ion Exchange

Let us find equations for internal-diffusion kinetics of adsorption (ion exchange) of a dissolved substance or gas for the case of an adsorbing medium consisting of individual porous particles alike in size and shape.

1. Sphere. Diffusion in a sphere is described by a system of equations of material balance for the substance diffusing through pores:

$$\frac{\partial q}{\partial t} + \frac{\partial C}{\partial t} = D \left(\frac{\partial^2 C}{\partial r^2} + \frac{2}{r} \frac{\partial C}{\partial r} \right), \tag{4.27}$$

and equations of adsorption isotherms:

$$q = \varphi(C), \tag{4.28}$$

where D is the diffusion coefficient of the substance in pores of the adsorbent.

The expression in parentheses is the notation of the Laplace operator in spherical coordinates taking the symmetry of the problem into account. Differentiating Eq. (4.28) according to time, and substituting in Eq. (4.27), we obtain

$$\frac{\partial C}{\partial t} = \frac{D}{1 + \varphi'(C)} \left(\frac{\partial^2 C}{\partial r^2} + \frac{2}{r} \frac{\partial C}{\partial r} \right), \tag{4.29}$$

where

$$\varphi'(C) = \left(\frac{\partial q}{\partial C} \right)_t.$$

Let us assume that the grains of the adsorbent rock, initially free of the substance, are placed in solution or gas of constant concentration C_0. The initial and boundary conditions are written in the form

$$C(r_0,\ t) = C_0, \quad C(r,\ 0) = 0, \tag{4.30}$$

$$\left[\frac{\partial C(r,\ t)}{\partial r}\right]_{r=0} = 0 \quad \text{(condition of symmetry)},$$

where r_0 is the radius of a grain.

A solution of Eq. (4.29) in analytical form may be obtained for the linear adsorption isotherm (3.10) by substituting in Eq. (4.29) u = rC. As a result [6], we obtain

$$\frac{C(r,\ t)}{C_0} = 1 - \frac{2}{\pi}\sum_{n=1}^{\infty}\frac{(-1)^{n+1}}{n}\frac{r_0}{r}\sin\left(\frac{n\pi r}{r_0}\right)\exp\left(-\frac{n^2\pi^2 D_{\text{ef}}}{r_0^2}t\right), \tag{4.31}$$

where the effective coefficient of interior diffusion is

$$D_{\text{ef}} = \frac{D}{1+K}. \tag{4.32}$$

An expression for the mean concentration in a grain (according to the definition of mean) is found from

$$\bar{C}(t) = \frac{3}{r_0^3}\int_0^{r_0} r^2 C(r,\ t)\,dr \tag{4.33}$$

and is written [5, 6, 8] in the form

$$\frac{\overline{C}(t)}{C_0} = 1 - \frac{6}{\pi^2}\sum_{n=1}^{\infty}\frac{1}{n^2}\exp\left[-\frac{n^2\pi^2 D_{\text{ef}}t}{r_0^2}\right]. \tag{4.34}$$

The amount of substance adsorbed by the grain is found in accordance with Eq. (3.11) and is equal to

$$\frac{q(t)}{q_0} = 1 - \frac{6}{\pi^2}\sum_{n=1}^{\infty}\frac{1}{n^2}\exp\left[-\frac{n^2\pi^2 D_{\text{ef}}}{r_0^2}t\right], \tag{4.35}$$

where $q_0 = KC_0$.

Equation (4.35) gives the change in amount of substance adsorbed by the grain in time $q = q(t)$; i.e., it is the equation of internal-diffusion kinetics of adsorption from a solution of steady concentration. Tables of values of the function $q(t)/q_0$ for different $\pi^2 D_{\text{ef}}t/r_0^2$ are given, for example, in the paper of Boyd, Adamson, and Meyers [8]. By using these tables and the experimental values of $q(t)/q_0$, obtained from a study of adsorption from a large volume of solution or from a flowing, constantly renewed solution [which is necessary in order to fulfill the boundary conditions (4.30)], it is possible to find the diffusion coefficient D_{ef} of the substance in the body of the grain (for methods of studying the kinetics, see § 33).

The equations of internal-diffusion kinetics of adsorption (4.34) and (4.35) are fulfilled when the concentration C_0 of external solution (pressure of gas) is low, and the adsorption isotherm is linear (see Chap. 3). An analytical solution of the problem of diffusion of the adsorbate substance through the pores of spherical grains of a rock may be obtained also for a different extreme case, when the adsorption capacity of the rock is small but the concentration is large. Let

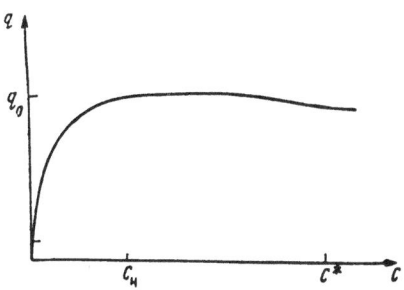

Fig. 14. Langmuir adsorption
isotherm.

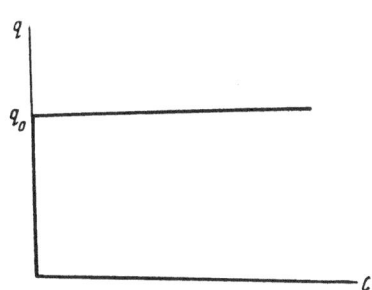

Fig. 15. Approximate rectilinear
adsorption isotherm.

$C_0 \gg C_s$, where C_s is the concentration of the solution at which the amount of adsorbed substance reaches the limiting value $q = q_0$ (Fig. 14). Then the adsorption isotherm will be approximately represented by a straight line (Fig. 15). In this

$$q \approx q_0 = \text{const} \quad \text{for all} \quad C > 0. \tag{4.36}$$

The equation of material balance (4.27) takes on the form

$$\frac{\partial C}{\partial t} = D \left(\frac{\partial^2 C}{\partial r^2} + \frac{2}{r} \frac{\partial C}{\partial r} \right). \tag{4.37}$$

Solving Eq. (4.37) with conditions (4.30) and converting by means of Eq. (4.33) from $C(r,t)$ to the mean concentration $\overline{C}(t)$ in the grain, we obtain

$$\frac{\overline{C}(t)}{C_0} = 1 - \frac{6}{\pi^2} \sum_{n=1}^{\infty} \frac{1}{n^2} \exp\left(-\frac{n^2 \pi^2 D}{r_0^2} t \right). \tag{4.38}$$

Thus, the adsorption process with time, in the investigated case, is determined by Eqs. (4.36) and (4.38).

If adsorption takes place from a solution limited by constant volume, and the concentration changes with time [$S(r_0,t) \neq$ const], the occurrence of the process in time is characterized by the following equation [5]:

$$\frac{q(t)}{q_0} = 1 - \frac{2}{3W} \sum_{n=1}^{\infty} \frac{1}{1 - \frac{S_n^2}{9W(W+1)}} \exp\left(-\frac{S_n^2 D_{\text{ef}}}{r_0^2} t \right), \tag{4.39}$$

where $W = CV/C_0 V_0$, S_n is the root of the transcendental equation $S_n \cot S_n = 1 + S_n^2/3W$, C_0 is the initial concentration of the adsorbate substance, and V_0 is the volume of the adsorbent brought into contact with volume V of the solution.

2. C y l i n d e r . The diffusion of dissolved substance or gas through pores of adsorbent grains having the form of a cylinder is characterized by the following equation:

$$\frac{\partial C}{\partial t} = \frac{D}{1 + \varphi'(C)} \left(\frac{\partial^2 C}{\partial r^2} + \frac{1}{r} \frac{\partial C}{\partial r} + \frac{\partial^2 C}{\partial z^2} \right), \tag{4.40}$$

where r is the radius vector and z is the coordinate along the axis of the cylinder.

The expression in parentheses is the Laplace operator acting on the function $C(r,z,t)$ in cylindrical coordinates (see Chap. 2), taking into account the axial symmetry of the problem.

Let grains of rock having the form of a cylinder with radius r_0 and length l, free of the adsorbate substance, be placed at time $t = 0$ in a solution of gas of constant concentration C_0. The initial and boundary conditions of the processes are

$$C(r_0,\ z,\ t) = C_0; \quad C(r,\ \pm l/2,\ t) = C_0; \quad C(r,\ z,\ 0) = 0;$$

$$\frac{\partial C(0,\ z,\ t)}{\partial r} = 0; \quad \frac{\partial C(r,\ 0,\ t)}{\partial z} = 0 \quad \text{(condition of symmetry)}. \tag{4.41}$$

Solving the system of equation (4.40) and (4.41) for the linear adsorption isotherm (3.11) and converting from $C(r,z,t)$ to the mean concentration in the cylinder, we finally obtain [6]

$$\frac{q(t)}{q_0} = 1 - \frac{32}{\pi^2} \sum_{n=1}^{\infty} \sum_{m=1}^{\infty} \frac{1}{\mu_n^2 (2m-1)^2} \exp\left[-\left(\frac{\mu_n^2}{r_0^2} + \frac{(2m-1)^2 \pi^2}{l^2} \right) D_{\mathrm{ef}} t \right], \tag{4.42}$$

where $q_0 = KC_0$, and μ is the root of a Bessel function of the first kind and order n ($\mu = 2.405$, $\mu_2 = 5.520, \ldots$).

Expression (4.42) is an equation of internal–diffusion kinetics of adsorption on rocks of cylindrical form from a solution of constant concentration. Timofeev [6] provided curves for the time dependence of $q(t)/q_0$ for adsorbent grains with different ratios of l/r_0. By using these curves, it is possible to find the coefficient of internal diffusion D_{ef} (with adsorption on cylindrical grains).

With adsorption from constant volume, the number of adsorbed grains at any instant of time is given by the following expression [6]:

$$\frac{q}{q_0} = 1 - \sum_{n=1}^{\infty} \frac{4\lambda(1+\lambda)}{4(1+\lambda) + \lambda^2 q_n^2} \exp\left[-\frac{q_n^2 D_{\mathrm{ef}} t}{r_0^2} \right], \tag{4.43}$$

where $\lambda = 2V/(1 + K)V_0$, q_n is the root of the characteristic equation $2I_1(q) + \lambda q I_0 = 0$; and I_0 and I_1 are Bessel functions of the first kind.

The roots q_n have been given by Timofeev [6].

The equation of adsorption kinetics on cylindrical grains from concentrated solutions ($C_0 \gg C_s$) may be derived in a manner similar to Eqs. (4.36) and (4.38).

From the above discussion it may be seen that equations of internal–diffusion adsorption kinetics are determined by the shape of the grains and are very complex even for the simplest spherical grains. With these equations it has not been possible to obtain simple solutions of the problem of geochemical migration (Chap. 6). In Chap. 5 we shall derive a simple approximate equation of internal-diffusion adsorption kinetics that will agree with experimental data.

§ 32. The Effect of Temperature on Reaction Rate

The rate of chemical reaction increases with rise in temperature. According to the classic views set forth by Arrhenius, not all the molecules enter into the reaction, but only the "active" molecules, i.e., those possessing greater energy than the mean energy of the molecules at the given temperature. With rise in temperature, the number of "active" molecules increases. Consequently, the reaction rate increases. The temperature dependence of the constant of reaction rate is given by the equation of Arrhenius [1]:

$$K = Ce^{-E/RT},\qquad(4.44)$$

where R is the universal gas constant, C the pre-exponential factor, being constant for a given reaction, and E the activation energy.

The activation energy is the excess energy above the average energy of the molecules at a given temperature, possessed by the "active" molecules [1]. The value of the activation energy for chemical reactions exceeds 20 kcal/mole. Equation (4.44) in coordinates (log K, 1/T) is represented by a straight line.

The views developed by Arrhenius for chemical reactions has proved to be applicable to a number of physicochemical processes, particularly to diffusion. Thus, for the coefficient of internal-diffusion of adsorbate gas, we have [9]

$$D = Ae^{-E/RT},\qquad(4.45)$$

where A is a constant value and E the activation energy.

Since the rate of adsorption and ion exchange is controlled by diffusion, the effect of temperature on the rate constants of these processes is determined by the temperature dependence of the coefficients of external and internal diffusion and by the functional relation between the constants of adsorption rates and diffusion coefficients. The temperature dependence of the coefficient of volume diffusion was examined in Chap. 2. The connection between the constants of adsorption and ion-exchange rates and the coefficients of internal and external diffusion will be examined in Chap. 5. By using these data we may determine the form of the theoretical temperature dependence of the constants of adsorption and ion-exchange rates.

It is known that the activation energy of diffusion is low (5-10 kcal/mole). It is considerably lower than the activation energy of chemical reaction. In accordance with Eq. (4.44), the constant of the rate of the heterogeneous process taking place in the diffusion region increases with rise in temperature more slowly than the constant of the rate of the same process in the kinetic region.

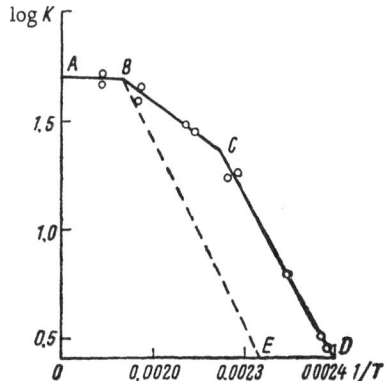

Figure 16 shows the relationship between rate constants and reciprocal temperature (1/T) for the catalytic reaction of acetone oxidized to manganese peroxide [1]. On the segment of the curve AB, the dependence of log K on 1/T is insignificant, corresponding to the diffusion region of reaction. Segment CD corresponds to the kinetic region, where the rate of reaction depends strongly on temperature, and the BC segment is transitional. Ordinarily the transition from one region to the other is smooth, so that the function log K = f(1/T) obtained by experiment has smoother continuity than the curve shown in Fig. 16.

Fig. 16. Dependence of the logarithm of apparent rate constants on the reciprocal temperature for the oxidation of acetylene to manganese peroxide, catalyzed by silver.

On the basis of the indicated differences in activation energy, we may determine in what region — diffusion or kinetic — the heterogeneous chemical reaction takes place. For this, it is necessary to measure the reaction rate at different temperatures and to determine the activation energy from these data. The order of these values (5–10 kcal/mole or ≥ 20 kcal/mole) furnishes an answer to the question concerning the range in which the reaction takes place.

§ 33. Methods of Studying the Kinetics of Heterogeneous Processes

An important question confronting one in making kinetic experiments concerns the method of making the investigation. At present there are several experimental methods for measuring reaction rates. Let us examine the fundamental aspects of the principal existing methods, their comparative advantages and defects.

The Static Method.   The essence of this method is the production of a reaction in a closed space with definite quantities of initial substances. In this method we study the change in concentration or partial pressure of the components of the reaction as this depends on time, total pressure, initial composition, and the like.

The principal advantage of the static method is the possibility of working with very small amounts of initial material, in obtaining the entire kinetic curve in a single experiment, and in the high sensitivity and accuracy of the measurements. One defect of the method is the difficulty facing the investigator in treating the experimental results. This follows from the fact that the concentrations of the reactants do not remain constant during the course of the kinetic experiment. As a consequence, the differential equations of rate [analogous to (4.10), (4.15), and (4.17)] are not integrated. In a number of particular cases, however, integration may be carried out.

The Dynamic Method.   This method involves the passing of liquid or gaseous reactants of constant concentration through a layer of adsorbent, catalyst, or rock at a constant rate. The change with time of the concentrations of components moving through the layer, as they emerge from the layer, serves to describe the rate of the heterogeneous reaction.

The dynamic experiment must be made in a regime of ideal displacement ("piston discharge") [10]; i.e., the particles of the mobile phase present in the layer must be completely displaced by the entering flow of particles without mixing. Such a regime is possible when there is little longitudinal diffusion of particles in the system (along the direction of flow). It is especially difficult to eliminate the effect of longitudinal diffusion on the results of measurements when the kinetics is studied at low flow rates (rates that are important in geochemical processes). In Chap. 5 we shall describe the design of an instrument for studying the kinetics of ion exchange in rocks, in which longitudinal diffusion plays no significant role.

The principal advantages of the dynamic method over the static method are discussed in the next few sentences. In order to describe the dynamics of adsorption or filtration of a substance in rocks , it is necessary, in the general case, to know the kinetic equations of the heterogeneous processes in the current (Chaps. 1 and 6). These equations are found experimentally from the dynamic experiment. The kinetic data obtained in a static experiment may be used for describing the filtration of substances interacting with the medium only when the process takes place in the internal–diffusion or kinetic regions.

Experimental results obtained by the dynamic method are more easily processed because of the constant concentration of substance supplied to the reaction zone, such as the layer of adsorbent. The condition of constant concentration is fulfilled when the layer of adsorbent is very thin. If the investigated adsorbent consists of individual granules, the measurements of rate must be made in a layer the thickness of one grain [3]. This condition cannot be fulfilled when the adsorption capacity of the rock is small. Kinetics is normally studied in a stream flowing through a thin layer of the adsorbent, the thickness of which is several times the diameter of a grain. In such experiments the concentration dependence curve of solution emerging from the thin layer is not the true kinetic curve of the heterogeneous process. It is a dynamic out-

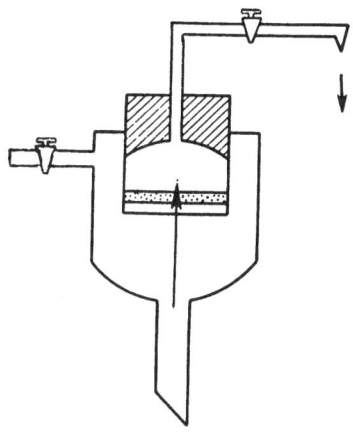

Fig. 17. A cell for studying ion exchange from a flow of solution.

put curve of the thin layer. It cannot therefore be described directly by corresponding kinetic equations. The discovery of kinetic relations of dependence by the type of dynamic curves of the process in its general form is a complex problem. It will be discussed in Chap. 6.

A number of authors [4, 11] used a simple apparatus, illustrated in Fig. 17, to measure the rate of ion exchange. A weighed sample of ionite is placed on a filter, and the solution is passed upward through it (as indicated by the arrow).

There are also circulation and flow-circulation methods for studying the kinetics of gas and vapor adsorption, having a number of advantages over the methods discussed.

§ 34. Experimental Data on the Kinetics of Heterogeneous Processes in Rocks

In Chap. 6 we will show that for a quantitative description of the geochemical migration of dissolved substances it is necessary to know the constants of the interaction rate between the solutions and rocks. Consequently, a study of the kinetics of heterogeneous processes in rocks has important significance. However, most investigations on the kinetics of heterogeneous processes in synthetic adsorbents and exchangers have given very little attention to the kinetics of adsorption, ion exchange, and chemical reactions in rocks.

An investigation on the kinetics of the adsorption of phosphate by soils — a process very important in agrochemistry — has been made by Fokin [12]. His studies were made on typical heavy clayey podsol, gleyish, with particle sizes less than 1 mm. The adsorption kinetics of potassium phosphate with concentrations of $1N$ and $0.1N$ was studied under static conditions by means of radioactive tracers (the phosphorous tracer P^{32}). It was found that most of the phosphorous was adsorbed after 5-10 minutes. After that interval, adsorption took place slowly, and equilibrium was established in 5-7 days.

Fokin gives no unique explanation of the observed picture, but he makes two suggestions.

1.　At first a more rapid process of reversible ion exchange takes place. Then slower irreversible adsorption takes over. With this explanation, it is assumed that adsorption takes place in the kinetic region.

2.　The slow adsorption is due to the low rate of diffusion of phosphate ions to remote adsorption sites; i.e., adsorption takes place in the internal-diffusion range. Fokin, however, for treating the results of measurements, did not use equations available from the literature for internal-diffusion kinetics of adsorption (ion exchange) from a solution of limited volume. Therefore, he was unable to obtain a unique explanation of the nature of the observed phenomena.

If reversible ion exchange takes place in the soil at the very initial moment, then, without doubt, it takes place in the diffusion region of kinetics. It is possible that, beginning at a certain time, the process takes place in the kinetic region (if the mechanism of adsorption undergoes a change and acquires an irreversible character, and the rate of this process becomes less than the diffusion rate).

We have investigated the kinetics of the interaction between solutions of $CuCl_2$ and HCl and bentonite in order to apply the results to a quantitative description of the filtration of solutions in bentonite. Rate measurements were made by the dynamic method, and the results of these measurements are given in the following chapter.

LITERATURE CITED

1. Panchenkov, G. M., and Lebedev, V. P., Chemical Kinetics and Catalysis [in Russian], Izd. MGU (1962).
2. Levich, V. G., Physicochemical Hydrodynamics [in Russian], Izd. Fiz.-Mat. Lit. (1959).
3. Golubev, V. S., and Panchenkov, G. M., Zh. Fiz. Khim., Vol. 36, p. 2271 (1962).
4. Golubev, V. S., Dissertation, Moscow University (1964).
5. Helfferich, F., Ion Exchange, McGraw-Hill, New York (1962).
6. Timofeev, D. P., The Kinetics of Sorption [in Russian], Izd. AN SSSR (1962).
7. Holm, L. W., J. Chem. Phys., Vol. 22, p. 1132 (1954).
8. Boyd, G. E., Adamson, A. W., and Myers, L. S., J. Amer. Chem. Soc., Vol. 69, p. 2836 (1947).
9. Glasstone, S., Laidler, K. J., and Eyring, H., The Theory of Rate Processes, McGraw-Hill Book Company, New York (1941).
10. Kiperman, S. L., Introduction to the Kinetics of Heterogeneous Catalytic Reactions [in Russian], Izd. Nauka (1964).
11. Gorshkov, V. I., Panchenkov G. M., and Ivanova, T. V., Zh. Fiz. Khim., Vol. 36, p. 1690 (1962).
12. Fokin, A., Dokl. Timiryazevskoi S. -Kh. Akademii, No. 89, p. 290 (1963).

CHAPTER 5

THE KINETICS OF ADSORPTION, ION EXCHANGE, AND CHEMICAL REACTION IN A CURRENT

§ 35. The Kinetics of Adsorption and Ion Exchange in the External Diffusion Region

Since a current of solution is present during migration induced by filtration, to solve the problem of geochemical migration it is necessary to know the kinetic equations of adsorption, ion exchange, and chemical reactions in a current. Derivation of theoretical equations describing the kinetics of interaction between solutions and host rocks in a current is most complex when the process takes place in the intergranular-diffusion region, where the rate depends on the hydrodynamics of flow. We shall point out how these equations might be found on the basis of sorption (ion exchange) on the spherical grain of a mineral in laminar flow [1-2].

We shall examine the following problem. Through a layer of rock or soil consisting of spherical grains of radius r_0, a liquid solution of substances being adsorbed is passed at a constant rate u. Conditions are thus created (a thin layer, rather high rate of supplying the substance) that the concentration of the solution may be considered constant at any instant of time at any point distant from the grains. This is considered an equilibrium situation then; equilibrium is established instantly both on the surface of the grains and within them.

It is easy to see that when adsorption takes place from a flow of solution, two mechanisms act to convey the adsorbed substance to the grain: diffusion and supply of the substance by the current. At different ratios of rates between intergranular diffusion and flow, adsorption may take place in the following kinetic regions: 1) the rate of supply of the substance being adsorbed to the grain by diffusion is much greater than the rate of supply by flow; then the adsorption rate (ion-exchange rate) is determined by flow (for more details, see § 38); 2) the rate of diffusion transfer is less than or near the rate of supply by flow, in which case adsorption (ion exchange) takes place in the intergranular-diffusion region.

Let us derive the equation for intergranular-diffusion kinetics of adsorption. We shall assume that the regime of solution flow is laminar. Then all the liquid may be broken down into two zones: the zone of constant concentration of solution far from the surface of the grains and the zone near this surface where the concentration changes because of diffusion and adsorption. The second zone is somewhat different from the boundary diffusion layer suggested by Nernst. We find the law of concentration change in this boundary diffusion layer as a function of distance from the surface of the grain.

The equation of material balance (1.17) for an arbitrary volume V containing a mineral particle is complex, and its solution in the general form has not been obtained. We shall simplify the problem by applying the principle of quasi-stationary diffusion [2]. We shall assume that the change in concentration C (in the diffusion layer) and q (in the grain) because of adsorption takes place at a much slower rate than the leveling of concentration in the bulk solution of the adsorbent layer by means of diffusion:

$$\left|\frac{\partial C}{\partial t}\right| \ll |D\,\Delta C|; \quad \left|\frac{\partial q}{\partial t}\right| \ll |D\,\Delta C|. \tag{5.1}$$

Let us consider when these inequalities are valid. Since the leveling of the concentration of adsorbate substance by diffusion takes place at a distance $\delta \approx 10^{-2}$-10^{-3} cm [2-3], where δ is the average thickness of the Nernst diffusion layer, the time τ necessary for the substance to diffuse to the distance δ is

$$\tau = \frac{\delta^2}{D} \approx \frac{(10^{-2}-10^{-3})\ \mathrm{cm}^2}{10^{-5}\ \mathrm{cm}^2/\mathrm{sec}} = (0.1-10)\ \mathrm{sec}\ . \tag{5.2}$$

Expression (5.2) follows from Einstein's law of diffusion. The diffusion coefficient of the substance in the bulk solution is $D \approx 10^{-5}$ cm^2 sec.

If after time τ the concentration of external solution C has not been appreciably changed, then conditions (5.1) are fulfilled.

Then the problem may be considered quasi-stationary. In practice, the inequalities (5.1) are generally fulfilled, since the time of the kinetic experiment, during which the changes in concentrations C and q take place, is much greater than τ.

The equation of material balance (1.17) for the quasi-stationary conditions is written in spherical coordinates, taking the inequalities (5.1) into account in the form

$$u_r \frac{\partial C}{\partial r} + \frac{u_0}{r}\frac{\partial C}{\partial \theta} = D\left(\frac{\partial^2 C}{\partial r^2} + \frac{2}{r}\frac{\partial C}{\partial r}\right), \tag{5.3}$$

where u_r is the radial component of flow rate, and u_θ is the tangential component of $\vec{u}$. The boundary conditions for the investigated process will

$$\begin{cases} C = f(q), & r = r_0 \quad \text{(equilibrium at the surface of the grain)} \\ C = C_0, & r \to \infty \quad \text{(far from the surface of a particle)} \\ C = C_0, & r = r_0, \quad \theta = 0 \ \text{(the reference origin of angle } \theta) \end{cases} \tag{5.4}$$

The third condition means that the concentration of the solution at the point where the current strikes the particle ($r = r_0$, $\theta = 0$) coincides with the concentration in the bulk solution, since the impinging current has not been impoverished by diffusion.

Equation (5.3) is integrated for boundary conditions (5.4), taking into account the approximate expression for determining rate near the surface of a sphere in a stationary current [3,4]. In doing this, Eq. (5.3) reduces to an equation of the type for thermal conductivity [2, 3]. Integration of the latter is much simpler than integration of Eq. (5.3); it has been shown in the book of Levich [3]. By assuming that the thickness of the layer in which a sharp change in concentration of the adsorbate substance takes place is small as compared to a particle size, integration of Eq. (5.3), with conditions (5.4), gives the following concentration distribution in the diffusion layer [1-3]:

$$C(z,\ t) \approx \frac{C_0 - f(q)}{1.16} \int_0^z e^{-4.\,9z^3}\,dz + f(q), \tag{5.5}$$

where $z = z(r,\ \theta;\ D,\ u)$ are new variables, introduced to simplify the integration of Eq. (5.3).

To find the flow of adsorbate substance to a particle, we use Fick's first law. Differentiating Eq. (5.5) and converting from z to the variables $(r,\ \theta)$, we obtain the following expression for flow to the surface of a particle [1, 2]:

$$j_D = D \left(\frac{\partial C}{\partial z} \right)_{z_0 = z\,(r_0,\,\theta,\,D,\,u)} = \frac{D}{1.16} \sqrt{\frac{3u}{4Dr_0^2}} \; \frac{\sin \theta}{\left(\theta - \frac{\sin 2\theta}{2} \right)^{1/3}} \; [C_0 - f(q)], \qquad (5.6)$$

Let us find the change that interests us: the change in time of amount of substance adsorbed by a particle dq/dt in reference to unit volume. The full current of substance striking a particle is

$$I = \int_{(S)} j_D \, dS = 2\pi r_0^2 \int_0^\pi j \sin \theta \, d\theta \approx 7.9 D^{2/3} u^{1/3} r_0^{4/3} [C_0 - f(q)]. \qquad (5.7)$$

Under the conditions that all the substance approaching the surface of a particle is adsorbed, we obtain the following expression:

$$\frac{dq}{dt} = \frac{I}{4/3\pi r_0^3} = 1.9 D^{2/3} u^{1/3} r_0^{-5/3} [C_0 - f(q)] = \gamma [C_0 - f(q)], \qquad (5.8)$$

where γ is the kinetic coefficient of external diffusion, depending on flow rate, diffusion coefficient, and size of adsorbent particles. Equation (5.8) describes the occurrence of adsorption (ion exchange) in time at a fixed mineral grain, and it is the equation of intergranular diffusion kinetics of adsorption (ion-exchange) equilibrium from a current. Solution of Eq. (5.8) for the linear isotherm $f(q) = q/K$ has the form

$$q(t) = KC_0 \left[1 - e^{-\frac{\gamma}{K} t} \right]. \qquad (5.9)$$

The equation of intergranular-diffusion kinetics of adsorption (ion exchange), derived on the basis of the Nernst diffusion layer, is written [5] in the form

$$\frac{dq}{dt} = 3Dr_0^{-1}\delta^{-1} [C_0 - f(q)]. \qquad (5.10)$$

In comparing Eqs. (5.8) and (5.10) it is seen that they differ in the factor before the concentration gradient $[C - f(q)]$. This is due to the fact that in Nernst's theory the hydrodynamics of flow is not taken into account. Instead, this theory rests on such an indeterminate parameter as the thickness of the diffusion layer, which cannot be computed but which nevertheless enters into the equation of external diffusion kinetics. It may be pointed out that, according to Eq. (5.6), the effective thickness of the diffusion layer used in Eq. (5.10) [1–3] is equal to

$$\delta = \frac{1.16 \left(\theta - \frac{\sin 2\theta}{2} \right)^{1/3}}{\sin \theta} \sqrt{\frac{4Dr_0^2}{3u}}. \qquad (5.11)$$

From expression (5.11) it is seen that the layer is asymmetrical. Consequently, it is meaningless to introduce a concept of thickness of the diffusion layer surrounding a given particle. The equation of external-diffusion adsorption (ion-exchange) kinetics in the presence of turbulent flow might be obtained in a similar manner [2].

The problem of adsorption of gas from a current in the intergranular-diffusion region is formulated analogously. However, with the adsorption of gases it has not been possible to obtain an equation of external-diffusion adsorption kinetics by the method described above, since in the solution of Eq. (5.3) one cannot use the approximate expression for rate of gas movement near the mineral particle [4]. Zhukhovitskii and others, using the qualitative dimensional theory [6], found for a regime of streaming gas transitional between laminar and turbulent, the following expression for the kinetic coefficient (rate constant) of external diffusion [7]:

$$\gamma \approx u^{1/2} (2r_0)^{-3/2} D^{1/2}.$$ (5.12)

Expression (5.12) was obtained on the basis of the equation of heat transfer to an infinite cylinder from the flow of an ideal liquid.

In the general case, the equation of external-diffusion adsorption (ion-exchange) kinetics from a flow of solution of constant concentration may be written as follows from the theory of mass transfer [8] in the form (4.25), where the constant of diffusion rate γ depends on the diffusion coefficient, the rate of flow, and the geometric parameters of the medium. No explicit form of γ in the general case is known.

Equations (4.25) and (5.8), derived above, may be used for describing sorption from a flow of solution of variable concentration $C \neq C_0 = $ const; $C = f(t)$ when intergranular-diffusion may be considered a quasi-stationary process. Diffusion is quasi-stationary if the concentration of the external solution changes slowly. Thus, if for time τ, from Eq. (5.2), the concentration of the exterior solution cannot be appreciably changed, the equation of intergranular-diffusion adsorption (ion-exchange) kinetics from a flow of solution of variable concentration C may be written in the form

$$\frac{\partial q}{\partial t} = \gamma \, [C - f(q)].$$ (5.13)

Equation (5.13) will be used in Chap. 6 for solving the problem of geochemical migration due to diffusion in the exterior-diffusion range.

§36. The Kinetics of Adsorption and Ion Exchange in the Internal Diffusion Region

Since the current flows by only the outer surface of the grains of a mineral, diffusion within the grain does not depend on the hydrodynamics of flow and may be described by the same equations as for the absence of flow [in particular, Eq. (4.25) for spherical mineral grains]. However equations of internal-diffusion kinetics of adsorption (ion exchange) (4.35) and (4.42), derived on the basis of Fick's second law, are so complex that simple analytical solutions due to filtration in the intragranular-diffusion region have not been obtained by means of these equations.

One of the authors of the present work, jointly with G. M. Panchenkov, derived an approximate equation for internal-diffusion kinetics of adsorption (ion exchange)[2,9], which is supported by experiment, as investigations have shown [2, 13].

Let us examine adsorption on a spherical mineral grain with radius r_0, taking place in the intragranular-diffusion region with the boundary conditions

$$C = C_0, \quad r \geqslant r_0.$$ (5.14)

We shall assume that the concentration distribution of the adsorbed substance for the size of grain is linear. This assumption is commonly made [11] when considering the kinetics of heterogeneous chemical reactions. As shown by computations made by Timofeev [12], the basic change in concentration of adsorbed substance in the adsorbent grain is linear, and only at relative concentrations $q/q_0 = q/KC_0 < 0.2$ does some deviation from linear behavior begin to appear. Then, for time $t \geq \tau$, where τ is the time required for the concentration front to reach the center of the grain, in according with Fick's first law, we have

$$\frac{dQ}{dt} = 4\pi r_0^2 D_{ef} \frac{q(r_0) - q(0)}{r_0} = 4\pi r_0 D_{ef} [q(r_0) - q(0)],$$ (5.15)

where Q is the amount of substance adsorbed by a grain at time t, q (r_0) and q (0) are the concentrations of adsorbed substance on the surface and inside the grain, respectively, and D_{ef} is the effective coefficient of intragranular diffusion.

Since adsorption equilibrium obtains on the surface of a grain, for the linear isotherm q $(r_0) = KC_0$. Taking this into account, in place of expression (5.15) we obtain

$$\frac{dQ}{dt} = 4\pi r_0 K D_{ef} \cdot \left[C_0 - \frac{q(0)}{K} \right].$$ (5.16)

Since there is a correlation between the values of q (C) and Q when the concentration distribution of grain thickness is linear [2, 9]:

$$q(0) = [Q - \pi r_0^3 q(r_0)] \frac{3}{\pi r_0^3},$$ (5.17)

by integrating Eq. (5.16) with the initial conditions t = τ, Q = Q_τ, we obtain the following equation of interior-diffusion adsorption (ion-exchange) kinetics [2, 9]:

$$-\ln\left(\frac{Q - Q_\infty}{Q_\tau - Q_\infty}\right) = \frac{12 D_{ef}}{r_0^2}(t - \tau),$$ (5.18)

where Q_∞ and Q_τ represent the amounts of substance adsorbed at equilibrium and at time τ.

An equation of intragranular-diffusion adsorption kinetics for times t ≤ τ was obtained by Panchenkov and others [11]. Investigation of this equation has shown [2], that it is analogous to Eq. (5.18). Therefore, for any time, the following equation is approximately fulfilled [2]:

$$\frac{dq}{dt} = \frac{12 D_{ef}}{r_0^2} K \left[C_0 - \frac{q}{K} \right].$$ (5.19)

By integrating (5.19) with the initial conditions

$$t = 0, \quad q = 0,$$ (5.20)

and the corresponding absence of adsorbed substance in the grain at time zero, we obtain the following equation of intragranular-diffusion adsorption (ion-exchange) kinetics:

$$q = KC_0 \left[1 - e^{\frac{-12 D_{ef}}{r_0^2} t} \right].$$ (5.21)

In generalizing the results we have obtained, it might be stated that the kinetics of adsorption (ion exchange) in the intragranular-diffusion region may be approximately described by Eq. (4.25) if by γ we mean the rate constant of internal diffusion. This constant, by analogy with Eq. (5.19), depends on the coefficient of interior diffusion, the geometric shapes and sizes of the mineral grains, and the adsorption capacity, but it does not depend on the hydrodynamics of the current.

When the concentration of dissolved substance changes with time, C ≠ C_0 = const, C = f (t), the equation of interior-diffusion adsorption (ion-exchange) kinetics may be described approximately by Eq. (5.13) (by analogy with the preceding paragraph), if the interior diffusion is a quasi-stationary process. The latter takes place if for the time

$$\tau = \frac{r_0^2}{D_{ef}} \approx 10^6 r_0^2 \ (\text{sec}), \tag{5.22}$$

necessary for the substance to diffuse to the center of the grain, the concentration of exterior solution does not change appreciably.

§ 37. The Kinetics of Adsorption and Ion Exchange with Simultaneous Consideration of Intergranular and Intragranular Diffusion

We shall derive the kinetic equation of adsorption (ion exchange) in the mixed diffusion region on a spherical grain of rock or soil with a grain radius of r_0 [2, 13]. Let the concentration of adsorbed substance far from a grain be constant and equal to C_0 for the course of the entire process. We shall examine adsorption equilibrium. Both on the surface and within the grain, equilibrium is established instantly. Since the rates of intergranular and intragranular diffusion are comparable, the rate of adsorption (ion exchange) will be limited both by diffusion of adsorbate substance to the grain surface and to diffusion within the grain.

Equilibrium at the surface is established instantly; therefore the condition of the following equation must be fulfilled for a linear isotherm, in accordance with Eq. (5.8):

$$\frac{dq}{dt} = 1.9 D^{2/3} u^{1/3} r_0^{-5/3} \left[C_0 - \frac{q(r_0)}{K} \right] = \gamma_1 \left[C_0 - \frac{q(r_0)}{K} \right]. \tag{5.23}$$

Since sorption takes placed in the mixed diffusion region, along with Eq. (5.23) the following equality must be satisfied:

$$\frac{dq}{dt} = \frac{\gamma_2}{K} [q(r_0) - q(0)], \tag{5.24}$$

where γ_2 is the rate constant of intragranular diffusion, which, in accordance with Eq. (5.19), is equal to

$$\gamma_2 = \frac{12 D_{ef} K}{r_0^2}. \tag{5.25}$$

Finding from Eq. (5.17) the relation between q ($q = Q / \frac{4}{3} \pi r_0^3$) and $q(0)$ and integrating Eqs. (5.23) and (5.24) with the initial conditions of (5.20), we obtain the following equation describing the rate of adsorption (ion exchange) in the mixed diffusion range:

$$-\ln\left(1 - \frac{q}{q_\infty}\right) = \frac{1}{K\left(\frac{1}{\gamma_1} + \frac{1}{\gamma_2}\right)} t. \tag{5.26}$$

Equation (5.26) may be used as an approximate description of adsorption (ion exchange) from a current on mineral grains of arbitrary shape and under any hydrodynamic conditions of flow. However, an explicit form of the kinetic coefficients γ_1 and γ_2 (intergranular and intragranular diffusion) in the general case cannot be found. Accordingly, the differential equation of adsorption (ion-exchange) rate from a current of solution of variable concentration in the mixed diffusion region for any isotherm has the approximate form

$$\frac{\partial q}{\partial t} = \gamma [C - f(q)], \tag{5.27}$$

where

$$\frac{1}{\gamma} = \frac{1}{\gamma_1} + \frac{1}{\gamma_2}, \tag{5.28}$$

if intergranular and intragranular diffusion may be considered quasi-stationary processes [2]. Thus, the reciprocal of the kinetic coefficient (rate constant) or diffusion in the mixed region is equal to the sum of the reciprocals of the rate constants of intergranular and intragranular diffusion (results similar to (4.24)).

§ 38. The Kinetics of Adsorption and Ion Exchange due to Flow

Adsorption (ion exchange) under dynamic conditions may take place, along with the above-described kinetic regions, in the region of so-called kinetics because of the presence of a current, when the adsorption rate is determined by the rate of flow [2, 14, 15].

We shall derive a kinetic equation for flow for the case when, in an adsorbent layer the thickness of a single grain, the adsorbate substance with concentration C_0 impinges at a constant rate u. On 1 cm^2 of the layer in unit time falls an amount of adsorbate material uC_0, and the amount $u\bar{C}$ departs, where $\bar{C}$ is the concentration of adsorbate substance emerging from the layer. The difference between the amount impinging on each square centimeter of surface and the emerging substance for the time dt is equal to the amount of substance dQ in a layer of V = 1 cm^2 d (d = average thickness of adsorbent grain):

$$dQ = u\,(C_0 - \bar{C})\,dt. \tag{5.29}$$

The amount of substance dq for the case of a linear isotherm is

$$dq = \frac{dQ}{d} \left(\frac{K}{1+K} \right). \tag{5.30}$$

Using Eq. (5.30), in place of Eq. (5.29) we obtain

$$\frac{dq}{dt} = \frac{uK}{d\,(1+K)}\,[C_0 - \bar{C}]. \tag{5.31}$$

Since, in the investigated kinetic region, diffusion retardation of adsorption is absent, the substance emerging from the layer is in equilibrium with the amount adsorbed by the given layer, so that $\bar{C} = q/K$. The kinetic equation (in view of flow) then has the form

$$\frac{dq}{dt} = \frac{uK}{d\,(1+K)} \left[C_0 - \frac{q}{K} \right]. \tag{5.32}$$

Let us examine the conditions for which adsorption takes place in the kinetic region because of flow. The average time necessary for molecules of the adsorbate substance to diffuse to the surface of a grain is

$$\tau_1 = \frac{\delta^2}{D}, \tag{5.33}$$

where δ is the averaged thickness of the Nernst diffusion layer. The time for a molecule to pass through the layer by flow is

$$\tau_2 = \frac{d}{u}. \tag{5.34}$$

If $\tau_1 \ll \tau_2$, equilibrium may be established in the layer, and adsorption takes place in the kinetic region because of flow. From Eqs. (5.33) and (5.34) it is seen that this takes place under the conditions

$$\frac{\delta^2}{d} \frac{u}{D} \ll 1. \tag{5.35}$$

The solution of Eq. (5.32) for the linear isotherm q = KC has the form

$$q = KC_0 \left(1 - e^{-\frac{u}{d(K+1)} t}\right). \tag{5.36}$$

Timofeev [12, 14] gives a solution of Eq. (5.32) of adsorption kinetics in the presence of flow for the adsorption isotherm of Dubinin and Radushkevich (4.8).

§ 39. Determination of the Diffusion Mechanism Controlling the Rate of Adsorption and Ion Exchange

For a quantitative treatment of the results of kinetic experiments on adsorption and ion exchange, including studies on rocks and soils, it is necessary to determine the kinetic region in which the process lies. Several methods are available for determining the diffusion mechanism that controls the adsorption (ion-exchange) rate. We shall pause on those that follows from (5.8), (5.19), and (5.27). From these equations it may be seen that the adsorption (ion-exchange) rate from a current depends in different ways on the flow rate u (depending on the region in which the process takes place. Experimental data have yielded the time dependent function $-\ln\left(1 - \frac{q(t)}{q_\infty}\right)$. In accordance with theory [Eqs. (5.9), (5.21), and (5.26)], this is a linear function. The slope of the line, as follows from Eqs. (5.9), (5.21), and (5.26), is γ/K. For determining the range in which the process takes place, it is necessary to take the kinetic adsorption (ion-exchange) curves q(t) for different flow rates u and plot the function of the kinetic coefficient as it depends on u. In the general case this dependence should have the form illustrated in Fig. 18. When $u > u_2$ the adsorption rate does not depend on flow rate. This is the intragranular region. The dependence of γ on u in the intergranular region differs from that for the mixed region, and this permits us to use a number of methods for delimiting these regions. Consequently, it is possible to determine the rate u below which adsorption lies in the intergranular region. Similarly, we may distinguish the kinetic region due to flow by using (5.32).

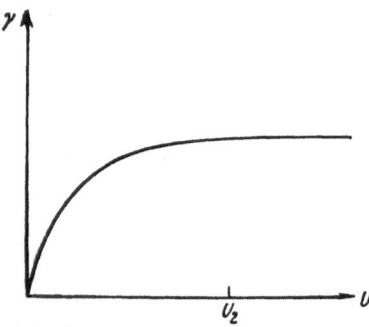

Fig. 18. Dependence of the kinetic adsorption coefficient γ on flow rate u.

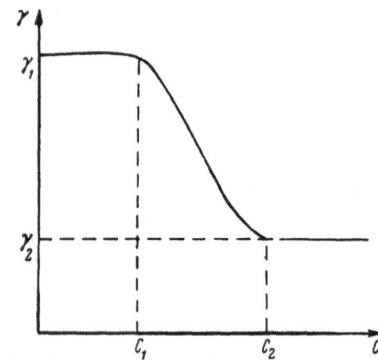

Fig. 19. Dependence of the kinetic ion-exchange coefficient γ on concentration of the solution C.

In the case of ion exchange it may be assumed [2] that the following method will also de-limit the kinetic region of the process. The dependence of the rate constant γ on the concentration C of the intergranular solution is plotted (Fig. 19). Since the range in which ion exchange takes place depends on C, then $\gamma = \gamma_1 = $ const ($0 \leq C \leq C_1$); this is the intergranular-diffusion range; γ smoothly grades from γ_1 to γ_2 in the mixed diffusion zone $C_1 \leq C \leq C_2$, and, lastly $\gamma = \gamma_2 = $ const in the intragranular-diffusion region ($C > C_2$). Thus, it has been possible to find the kinetic regions in which ion exchange of alkali metals with hydrogen on the cationites KU-1 and KU-2, and zeolites [2, 16] takes place.

§ 40. The Kinetics of Heterogeneous Chemical Reactions in a Current

The kinetic equation of heterogeneous chemical reaction taking place in a current, as for the kinetics of adsorption, may be obtained by solving the equation of material balance (1.17) for definite initial and boundary conditions. Assuming the concentration of the solution far from the reaction surface to be constant (because of flow), we obtained the first boundary condition

$$C = C_0, \quad z \to \infty, \tag{5.37}$$

where z is the distance to the reaction surface.

The other boundary conditions are more complex. They may be found from the following considerations [3]. The flow of substance to the surface of the mineral, in accordance with Fick's first law, is equal to

$$j_D = D \left(\frac{\partial C}{\partial z} \right)_{z=0}, \tag{5.38}$$

where the derivative is taken outward, normal to the surface. We shall assume that an irreversible reaction of n-th order takes place at the surface. The number of particles reacting per unit time per square centimeter of reaction surface, in accordance with the basic postulate of chemical kinetics (4.4), is

$$V = K \left(C_1 \right)^n, \tag{5.39}$$

where C_1 is the concentration on the reaction surface, and K is the rate constant of the reaction taking place on the surface.

Under stationary conditions, the number of particles arriving at the surface is equal to the number of reacting particles. We then obtain the following boundary condition:

$$\frac{D}{K} \left(\frac{\partial C}{\partial z} \right)_{z=0} - C_1^n = 0. \tag{5.40}$$

If $K \gg 1$, which corresponds to reaction in the diffusion region, the boundary condition acquires the form

$$C_1 = 0, \quad z = 0. \tag{5.41}$$

Condition (5.41) means that if the reaction rate is large, all particles arriving at the surface react instantaneously.

When $K \ll 1$, the process lies in the kinetic region, and the boundary condition is written

$$\left(\frac{\partial C}{\partial z}\right)_{z=0} = 0. \tag{5.42}$$

Condition (5.42) means that the concentration in the bulk solution does not change with time. The overall rate of the process is determined by the rate of chemical reaction:

$$V = K C_0^n. \tag{5.43}$$

From Eq. (5.43) it follows that if the heterogeneous process takes place in the kinetic region, its rate is described by the corresponding equations of formal kinetics.

Let us find equations of rate for a number of reactions taking place in the kinetic region. We shall take into account the fact that heterogeneous processes generally take place with no substantial change in volume of the host rock [17]. We may therefore use the determination of reaction rate in accordance with Eq. (4.3).

1. I r r e v e r s i b l e   R e a c t i o n   o f   t h e   S e c o n d   O r d e r . Let a solution A, flowing around grains of rocks, react with some component B in the rock, forming an insoluble compound D. This reaction may be represented schematically as follows:

$$A - B \xrightarrow{K'} D, \tag{5.44}$$

where A, B, and D are the designated substances. By q_0 we shall designate the concentration of the active component in the rock before beginning of the reaction, q the concentration of the substance obtained by the reaction, and K' the rate constant of the reaction.

In accordance with Eq. (4.4), we obtain the following kinetic equation:

$$\frac{\partial q}{\partial t} = K'C\,(q_0 - q). \tag{5.45}$$

2. R e v e r s i b l e   R e a c t i o n   o f   t h e   F i r s t   O r d e r . The reaction may be represented schematically as follows:

$$A \underset{K_2}{\overset{K_1}{\rightleftarrows}} B, \tag{5.46}$$

where K_1 and K_2 are the rate constants of the forward and reverse reactions, respectively.

The equation of reaction rate, in accordance with the basic postulate of chemical kinetics (Chap. 4), has the form

$$\frac{\partial q}{\partial t} = K_1 C - K_2 q = K_1 \left(C - \frac{q}{K_c}\right), \tag{5.47}$$

where $K_c = K_1/K_2$ is the equilibrium constant of reaction (5.46).

3. R e v e r s i b l e   R e a c t i o n   o f   t h e   S e c o n d   O r d e r . Let us consider reactions of the type

$$A + B \underset{K_2}{\overset{K_1}{\rightleftarrows}} D, \tag{5.48}$$

$$A + B \underset{K_2}{\overset{K_1}{\rightleftarrows}} D + E. \tag{5.49}$$

In reaction (5.48) the rate of the reverse process obeys the monomolecular law, but in reaction (5.49), the bimolecular law. Rate equations of these reactions, as may be readily seen, are written

$$\frac{\partial q}{\partial t} = K_1 C \left(q_0 - q \right) - K_2 q, \qquad (5.50)$$

$$\frac{\partial q}{\partial t} = K_1 C \left(q_0 - q \right) - K_2 q^2. \qquad (5.51)$$

4. Chemisorption. Frequently, chemisorption may be formally considered an irreversible reaction of the form

$$\begin{array}{c} \text{Adsorbable} \\ \text{substance} \end{array} + \begin{array}{c} \text{Free sites on} \\ \text{the adsorbent} \end{array} \xrightarrow{\beta} \begin{array}{c} \text{Adsorbed} \\ \text{substance} \end{array} \qquad (5.52)$$

where β is the rate constant of the reaction. If we consider the interaction between the adsorbed molecules, the kinetic equation of reaction (5.52) may be written [1, 2]

$$\frac{\partial q}{\partial t} = \beta C \varphi \left(q \right), \qquad (5.53)$$

where $\varphi \left(q \right)$ is a value that depends on the concentration of "free sites" on the adsorbent.

The function $\varphi \left(q \right)$ takes into account, in the most general case, interaction between the adsorbed molecules. Ordinarily, at saturation $q = q_0$ and $\partial q / \partial t = 0$. Therefore

$$\varphi \left(q_0 \right) = 0. \qquad (5.54)$$

In the last part of this section we shall derive equations of rate of the irreversible reaction (5.44) taking place in the diffusion region.

Through a layer of rock made up of spherical mineral grains of radius r_0, we shall filter a liquid solution of constant concentration C_0 at a constant rate, reacting on the grain surfaces with one of the rock components having initial concentration q_0. For time $t < \tau$, where τ is the time for which all of substance B in the rock is used up, the process is described by the equation of convective diffusion (5.3) with the conditions [comparable with Eq. (5.4)]

$$\begin{aligned} C &= 0, & r &= r_0, \\ C &= C_0, & r &\to \infty, \\ C &= C_0, & r &= r_0, & \theta = 0. \end{aligned} \qquad (5.55)$$

By analogy with Eq. (5.8) we obtain for the time $t < \tau$ the following kinetic equation:

$$\frac{dq}{dt} = 1.9 D^{2/3} u^{1/3} r_0^{-2/3} C_0, \ t < \tau, \qquad (5.56)$$

The solution of Eq. (5.56) has the form

$$q = 1.9 D^{2/3} u^{1/3} r_0^{-2/3} C_0 t, \ t < \tau. \qquad (5.57)$$

If the substance that formed as a result of the reaction does not retard the process, time τ is then equal to

$$\tau = \frac{q_0}{1.9 D^{2/3} u^{1/3} r_0^{-5/3} C_0} \cdot \tag{5.58}$$

Lastly, we obtain the following integral equation of the rate of reaction (5.44) taking place in the diffusion region:

$$q = \begin{cases} 1.9 D^{2/3} u^{1/3} r_0^{-5/3} C_0 t, & t \leqslant 0.53 D^{-2/3} u^{-1/3} r_0^{5/3} \frac{q_0}{C_0} \\ q_0, & t \geqslant 0.53 D^{-2/3} u^{-1/3} r_0^{5/3} \frac{q_0}{C_0}. \end{cases} \tag{5.59}$$

When the concentration of filtered substance changes with time, the problem is substantially complicated. However, if we consider diffusion to the surface of the grain as quasi-stationary, the differential equation of rate, by analogy with Eq. (5.13), is written in the form

$$\frac{\partial q}{\partial t} = \begin{cases} 1.9 D^{2/3} u^{1/3} r_0^{-5/3} C, & t < \tau, \\ 0, & t > \tau. \end{cases} \tag{5.60}$$

In the general case, the equation of rate of reaction (5.44) in the diffusion region may be written

$$\frac{\partial q}{\partial t} = \begin{cases} \gamma C, & t < \frac{q_0}{\gamma C_0}, \\ 0, & t > \frac{q_0}{\gamma C_0}, \end{cases} \tag{5.61}$$

where the kinetic coefficient γ depends on the diffusion coefficient, the flow rate, and so forth. It has not been possible to find an explicit expression for γ in the general case.

If reaction (5.44) takes place not only at the surface but also within the grains, the process may occur in the intragranular region of kinetics (if diffusion within the grains takes place more slowly than diffusion to the surface and to the reaction itself). The equation of intragranular kinetics of reaction (5.44), by analogy with the previous discussion, may be written in the form (5.61) if by γ we understand the rate constant of intragranular diffusion.

§ 41. Study of the Kinetics of Ion Exchange in Rocks from a Current

The kinetics of the adsorption of gases and vapors and the kinetics of ion exchange from a current has been investigated, as a rule, in synthetic adsorbents (see [2, 5, 7, 10, 12, 14–16, 18]).

Zabezhinskii, in investigating the adsorption rate of alcohol vapors on activated coal, found the following relation between the rate constant of intergranular diffusion and the parameters of the experiment [18]:

1) from experiments with individual grains:

$$\gamma \approx u^{0.4} (r_0)^{-1.6}; \tag{5.62}$$

2) from experiments with a layer of grains:

$$\gamma \approx u^{0.5} (r_0)^{-2.0}. \tag{5.63}$$

The data obtained do not agree with theoretical data (5.12), although they are close.

In studying the kinetics of the exchange of hydrogen ions for metal ions of groups I and II in the synthetic resins KU-1 and KU-2 and the exchange of metals in zeolite (inorganic exchanger) of type A, it has been possible to establish the fact [16] that the time dependence of $-\ln\left(1 - \dfrac{q}{q_\infty}\right)$ is approximately linear for any kinetic region in which the process takes place. These data, and also a number of other results [10], show that Eqs. (5.9), (5.21), and (5.26) may be used for describing the kinetics of adsorption and ion exchange in the intergranular, intra-granular, and mixed regions, respectively. As for interpreting the kinetic coefficient γ_1 in Eqs. (5.9) and (5.26) by means of Eq. (5.8), this question needs supplementary experimental verification.

As shown in Chap. 3, rocks in their adsorption and ion-exchange properties are similar to synthetic adsorbents. It should be expected that Eqs. (5.9), (5.21), and (5.26) would describe the kinetics of adsorption and ion exchange in rocks. However, the filtration rate of solutions in rocks is very low, and for such rates we have no experimental data on the kinetics of adsorption (ion exchange) in artificial adsorbents. Therefore, it would be interesting to study the kinetics of ion exchange in rocks at low flow rates.

For such a study it is probably impossible to use an experimental setup similar to that used for measurements on synthetic adsorbents (Fig. 17) for the following reasons. At very low flow rates, the arrival of the substance to an adsorbent layer the thickness of single grain will be determined not only by flow but also by longitudinal diffusion of the dissolved substance. This would lead to the impossibility of processing the experimental results by means of the kinetic equations (5.9), (5.21), and (5.26), since longitudinal diffusion was not taking into account in the derivation of these equations. The granules of the rock we selected — bentonite — were small, and its adsorption capacity was not very large. Therefore, the change in concentration of dissolved substance in the current because of adsorption may be determined rather accurately if the layer of rock with the thickness of a single grain covers a large area of the filter (3, in Fig. 20). With a large filter area, the paths traveled by the molecules of dissolved substance from the filter to the narrow tube at the outlet of the apparatus would differ, and this would lead to supplementary erosion of the concentration front at the outlet of the apparatus, and, consequently, the picture of interaction between solution and rock would be distorted (for more detail see Chap. 6).

Figure 20 illustrates schematically the experimental setup on which measurements were made on interaction rates of solutions of $CuCl_2$ and HCl with bentonite. In this apparatus, the undesirable effects of longitudinal diffusion and spreading of the concentration front have been eliminated to a considerable extent because of the different path lengths followed by the molecules of dissolved substance. On a Plexiglas screen, 3, 90 mm in diameter and 10 mm high, tightened from below by a dense network of polycaprolactam fibers, was placed a layer of pure quartz sand 5 mm high, with grain sizes of 0.2-0.4 mm. Granules of bentonite were sifted onto the sand through 0.2- and 0.3-mm sieves in a layer one grain thick. Quartz sand was piled on top of the bentonite layer. The screen was set on a glass funnel, 2, which was also filled with quartz sand. The layer of sand in the funnel was supported by a net of polycaprolactam fibers. In Fig. 20 it may be seen that the selected filter is characterized by the fact that the flow length over the entire filter is approximately the same.

The funnel, 2, was placed within a vessel. A stream of solution was directed on the funnel from above, sprayed on by means of a brush, 5. Droplets of solution that fell downward were combined on the surface of the filter, and the substance was carried to the rock by direct flow. This method of supplying substance to the rock eliminates longitudinal diffusion from the supply current and permits one to obtain low flow rates.

Experiments were carried out in the following way. Samples of bentonite, placed on the screen in the manner described above, were placed in the apparatus and moistened with water in order to prevent sorption of water from the flow of solution. The bentonite used in the

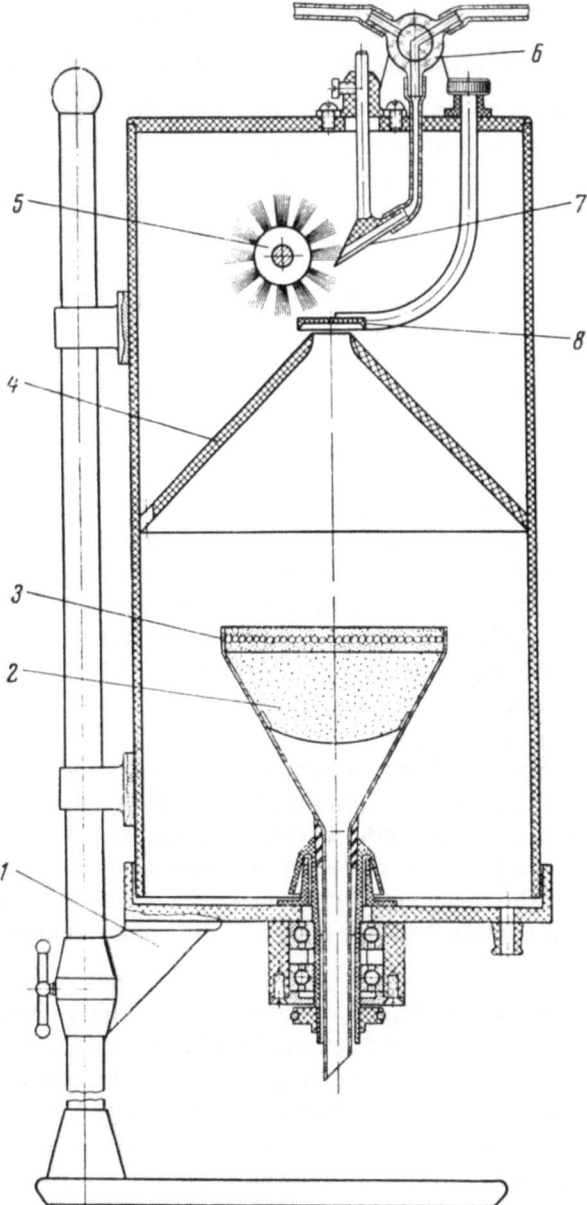

Fig. 20. Apparatus for studying rate of ion exchange in rocks with slow flow rates. 1) Drip
pan with attachment and outlet nozzle; 2) glass
funnel with quartz sand; 3) Plexiglas screen;
4) conical diaphragm; 5) brush for spraying
solution; 6) three-way tap; 7) channel for supplying solution to brush; 8) stop valve.

experiments was characterized by the following composition (in %): 66.98 SiO_2, 11.68 Al_2O_3,
2.04 Na_2O, 1.81 Fe_2O_3, 0.47 FeO, 1.35 MgO, 1.13 CaO, 0.16 TiO_2, 0.12 K_2O, a trace of MnO,
14.06 lost during roasting. A stream of $CuSO_4$ or HCl solution was directed onto the moistened
filter. Since natural solutions are ordinarily dilute, we used a solution of concentration 0.03 N.
At the outlet of the filter the solution was collected after different intervals of time in 15-ml

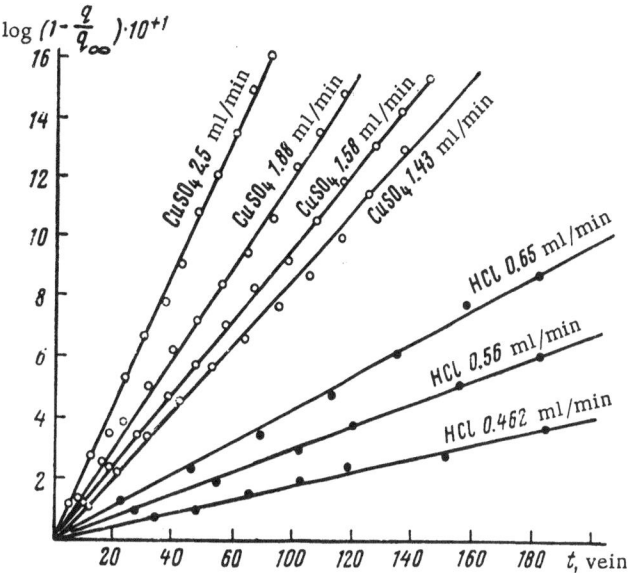

Fig. 21. Kinetic curves of ion exchange of HCl and CuSO₄ solutions on bentonite for different flow rates.

test tubes and were analyzed for the content of Cu^{2+} and H^+ ions by iodometric titration and by titration with alkali, respectively. The flow rate in different experiments ranged from 20 to 200 ml/hr, corresponding to linear velocities of the solution (if we assume the porosity of the bentonite layer to be 0.4) through the bentonite layer of u ≈ 0.015-0.15 cm/min. The rate constant of flow was controlled.

In Fig. 21 are shown, in the coordinates $\left[-\log\left(1 - \frac{q}{q_\infty}\right), t \right]$, the results of measuring the rate of ion exchange of Cu^{2+} and H^+ on bentonite. The exchange isotherms were first measured, and were found to be nearly linear. From Fig. 21 we see that the time dependence of $\log\left(1 - \frac{q}{q_\infty}\right)$ is linear, which agrees with the theoretical equations (5.9), (5.21), (5.26), and (5.36).

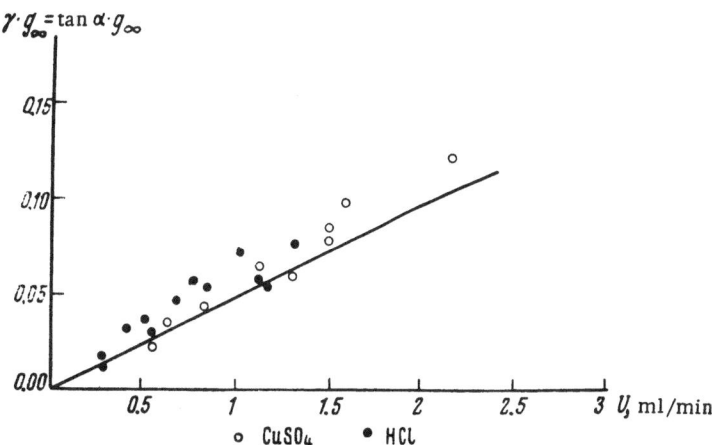

Fig. 22. Dependence of the kinetic coefficients of ion exchange for CuSO₄ and HCl on flow rate.

For determination of the kinetic region in which the process takes place, it is convenient to plot a graph of the dependence of γq_∞ on flow rate u (see Fig. 22). The kinetic coefficient γ is determined from the slope of the straight lines in Fig. 21. From Fig. 22 it follows that the dependence of γ on u is linear, and the values of γ for the exchange of Cu^{2+} and H^+ are the same, within the limits of measuring error. This indicates that at low flow rates u $\approx$ 0.015-0.15 cm/min and, also, when condition (5.35) is fulfilled, ion-exchange equilibrium in the bentonite layer may be established at any instant of time, and the rate of ion exchange is determined by the rate of supply of substance by the current. Actually, theory shows us that the linear dependence of γ on u obtains only for the case of kinetics determined by flow [Eq. (5.36)]. Furthermore, should the process take place not in the kinetic region because of flow, then the kinetic coefficients $\gamma_{Cu} \neq \gamma_H$ when u = const, since in any kinetic range $\gamma_{Cu} = f(D_{Cu})$; $\gamma_H = f(D_H)$, but the diffusion coefficients of copper D_{Cu} and hydrogen D_H are substantially different. The theoretical dependence $\gamma q_\infty = f(u)$, plotted in accordance with Eq. (5.36) (the solid line in Fig. 22), is in agreement with experimental results.

Thus, experimental data confirm the existence of the kinetic region at low flow rates (u $\approx$ 0.015-0.15 cm/min), in which the rate of ion exchange is determined by the supply rate of the substance by flow. The experimental method requires further improvement. One cause leading to divergence of the experimental points on Fig. 22 is found in the fact that at low flow rates it is not possible to maintain a strictly constant u during the course of the entire kinetic experiment.

LITERATURE CITED

1.   Golubev, V. S., and Panchenkov, G. M., Zh. Fiz. Khim., Vol. 36, p. 2271 (1962).
2.   Golubev, V. S., Dissertation, Moscow University (1964).
3.   Levich, V. G., Physicochemical Hydrodynamics [in Russian], Izd. Fiz.-Mat. Lit. (1959).
4.   Levich, V. G., Uspekhi Khimii, Vol. 34, p. 1846 (1965).
5.   Boyd, G. E., Adamson, A. W., and Myers, L. S., J. Am. Chem. Soc., Vol. 69, p. 2836 (1947).
6.   Gukhman, A. A., Introduction to the Theory of Similarity [in Russian], Izd. Vysshaya Shkola (1963).
7.   Zhukhovitskii, A. A., Zabezhinskii, Ya. L., and Venichkina, A., Zh. Fiz. Khim., Vol. 15, p. 174 (1941).
8.   Kafarov, V. V., Fundamentals of Mass Transfer [in Russian], Izd. Vysshaya Shkola (1962).
9.   Golubev, V. S., and Panchenkov, G. M., Zh. Fiz. Khim., Vol. 37, p. 1010 (1964).
10. Panchenkov, G. M., Golubev, V. S. Skoblo, V. A., and Rozen, I. A., Kinetics and Catalysis [in Russian], Tr. Moskovskogo Inst. Neftekhim. i Gazovoy Prom., No. 69, Izd. Khimiya (1967).
11. Panchenkov, G. M., Skoblo, V. A., and Zhorov, Yu. M., Izv. WZ, Neft' i Gaz, No. 1 (1961).
12. Timofeev, D. P., The Kinetics of Sorption [in Russian], Izd. AN SSSR (1962).
13. Golubev, V. S., and Panchenkov, G. M., Zh. Fiz. Khim., Vol. 38, p. 228 (1964).
14. Timofeev, D. P., Zh. Fiz. Khim., Vol. 29, p. 725 (1955).
15. Timofeev, D. P., Zh. Fiz. Khim., Vol. 32, p. 2005 (1958).
16. Golubev, V. S., Panchenkov, G. M., and Filimonov, V. G., Kinetics and Catalysis [in Russian], Tr. Mosk. Inst. Neftekhim. i Gazovoy Prom., No. 69, p. 151, Izd. Khimiya (1967).
17. Korzhinskii, D. S., The Physicochemical Basis for Analysis of the Paragenesis of Minerals [in Russian], Izd. AN SSSR (1957).
18. Zabezhinskii, Ya. L., Zh. Fiz. Khim., Vol. 17, p. 32 (1943).

CHAPTER 6

GEOCHEMICAL MIGRATION DUE TO FILTRATION AND DIFFUSION

§ 42. Diffusion in a Heterogeneous Medium with Consideration of Adsorption and Ion Exchange

In Chap. 2 it was shown how to solve the problem of diffusion in porous media in the absence of interaction between the diffusing substance and the medium. In nature, the diffusing substance is adsorbed by the rocks, exchanges ions with them, or enters into chemical reaction with the rocks.

Let us consider the diffusion of gas or single-component solution in a heterogeneous medium. The process in this case, as follows from Eqs. (1.9) and (1.17), is characterized by the following system:

$$\frac{\partial q}{\partial t} + \frac{\partial C}{\partial t} = \eta D \, \Delta C, \tag{6.1}$$

$$\frac{\partial q}{\partial t} = \varphi \, (C, \ q, \ K_i, \ D), \tag{6.2}$$

where D is the coefficient of diffusion in the bulk solution and η is the tortuosity factor.

The system (6.1) and (6.2) describes the diffusion of a substance dissolved in the liquid phase, with certain initial and boundary conditions, if the porous space of the medium is completely filled with water, i.e., if osmotic transfer of water is absent. An explicit form of the kinetic equation (6.2) depends not only on the method of interaction between the substance and the medium, but also on the physical characteristics of the medium (porosity, discreteness, fracturing, and so forth). The following distinctions may be made in regard to fracturing:

1. Diffusion in a continuous porous medium. Here we include diffusion along communicating microfractures in nonporous rocks. In this model of the medium, the pores between particles and within them are assumed to be alike, so that diffusion proceeds at the same rate between and within the grains.

2. Diffusion in a discrete porous medium, i.e., a medium consisting of separate particles (soil, clay minerals). In this case the size of the pores within particles differs from that of intergranular pores, as a consequence of which the diffusion coefficient in the particles (intragranular diffusion) differs from that in the space between them (intergranular).

3. Diffusion in a discrete medium so that the interaction with the solution takes place only on the surface of particles of the medium (glauconite).

We have formulated and found several solutions for each of the three cases, when the diffusing substance is adsorbed or exchanges ions with the rock. It is assumed that if diffusion takes place in the liquid phase the pores are completely filled with water.

   1. Diffusion in a Continuous Porous Medium. Numerous experimental studies on sorption and ion exchange have shown that diffusion ordinarily takes place much more slowly than the processes of sorption or ion exchange (see Chap. 4). As a consequence, the substance diffusing through the rocks is in equilibrium with the adsorbent walls of the pores at any instant of time. Then, as shown in Chap. 3, the concentration of adsorbed substance q is related to the concentration C in the bulk solution by the adsorption (ion-exchange) isotherm equation (4.28). Therefore, in place of Eq. (6.2), we may use the following equation, which is obtained by differentiating Eq. (4.28) with respect to time:

$$\frac{\partial q}{\partial t} = \frac{\partial q}{\partial C} \cdot \frac{\partial C}{\partial t} = \varphi'(C)\,\frac{\partial C}{\partial t}. \tag{6.3}$$

Substituting from expression (6.3) in (6.1) we obtain an equation characterizing the diffusion of dissolved substance in a continuous porous medium:

$$\frac{\partial C}{\partial t} = \frac{\eta D}{1 + \varphi'(C)}\,\Delta C \tag{6.4}$$

with definite initial and boundary conditions. Let us examine the solution of Eq. (6.4) for simple cases.

   If the adsorption (ion-exchange) isotherm is linear (the substance diffuses at low concentrations), then, in accordance with Eq. (3.11), in place of Eq. (6.4) we obtain

$$\frac{\partial C}{\partial t} = \frac{\eta D}{1 + K}\,\Delta C = D_a\,\Delta C, \tag{6.5}$$

where D_a is the diffusion coefficient in the adsorgent medium.

   A comparison of Eqs. (6.5) and (2.6) shows that, in the investigated example, the process is described by the equation of Fick's diffusion if the diffusion coefficient D is replaced by $D_a = D/(1+K)$. Since K is essentially positive, $D/(1+K) < D$. As follows from Eqs. (2.36), (2.37), and (2.51), adsorption (ion exchange) leads to a decrease in distance the substance migrates through the rock pores by a factor of $\sqrt{1+K}$ as compared with migration in the absence of sorption. For describing diffusion geochemical migration of substances in low concentrations we may use solutions of the problem of nonstationary diffusion [including Eqs. (2.33), (2.40), (2.45), (2.47), and (2.54)] if the diffusion coefficient D is replaced by $D_a = D/(1 + K)$.

   The problem of diffusion is fundamentally complicated when the adsorption isotherm is not linear. In this case, as we may see from Eq. (6.4), it is necessary to seek a solution of the equation of nonstationary diffusion with a diffusion coefficient depending on concentration. For example, for the Langmuir isotherm (3.10) Eq. (6.4) acquires the form

$$\frac{\partial C}{\partial t} = \frac{\eta D}{1 + \dfrac{a}{(1+bC)^2}}\,\Delta C. \tag{6.6}$$

   Solutions of Eq. (2.6) with variable diffusion coefficients have been made in a number of papers [1-6]. However, only in individual cases was it possible to obtain analytical solutions of the problem (normally in the form of infinite series). Most commonly the solutions were found by using various approximations or numerical methods, and were represented in the form of dimensionless graphs.

Let us consider a solution to the diffusion equation (6.4) in a semiinfinite body for the convex adsorption (3.9) of Dubinin and Radushkevich [5]. The system of equations (6.4) and (3.9) describes in particular the diffusion of vapors of organic substances in activated coal. Differentiating Eq. (3.9) according to C, substituting in (6.4), and neglecting the units in the denominator, we obtain for diffusion along the x axis

$$\frac{\partial C}{\partial t} = \frac{\eta D \beta^2 V C}{2 B W_0 T^2 \log \frac{C_s}{C} \exp\left[-\frac{BT^2}{\beta^2}\log^2\frac{C_s}{C}\right]} \cdot \frac{\partial^2 C}{\partial x^2}.$$

(6.7)

We shall seek a solution of Eq. (6.7) for the following initial and boundary conditions:

$$\begin{aligned} C(x, 0) &= 0, \qquad x > 0, \\ C(0, t) &= C_0. \end{aligned}$$

(6.8)

D. P. Timofeev simplified Eq. (6.7) in the following manner. The value of

$$F = \log \frac{C_s}{C} \exp\left[-\frac{BT^2}{\beta^2}\log^2\frac{C_s}{C}\right]$$

(6.9)

changes little with change in C, and, as a first approximation, we may assume it to be constant.

Then Eq. (6.7) takes on the form

$$\frac{\partial C}{\partial t} = HC \frac{\partial^2 C}{\partial x^2},$$

(6.10)

where

$$H = \frac{\eta D \beta^2 V}{2 B W_0 T^2 F}.$$

(6.11)

Timofeev further uses the approximation method of solving Eq. (6.10) on the basis of using the solution of Eq. (2.45) of nonstationary diffusion with constant diffusion coefficient. For this he split the adsorption isotherm into n segments (C_1, C_2, ..., C_i, ..., C_n with intervals $\Delta C = C_0/n$ and replaced it by a broken line (this may be done if ΔC is small). Then, in place of Eq. (6.10) we obtain a system of n equations:

$$\frac{\partial C_i}{\partial t} = D_i \frac{\partial^2 C_i}{\partial x^2}$$

$$(i = 1, 2, \ldots, n),$$

(6.12)

where the effective coefficient of diffusion for the i segment of the isotherm is

$$D_i = HC_i.$$

(6.13)

By designating the diffusion coefficient at i = 1 by D_0, as follows from Eq. (6.10)

$$D_0 = HC_0,$$

(6.14)

and by carrying out a number of transformations, Timofeev obtained n equations with constant diffusion coefficients:

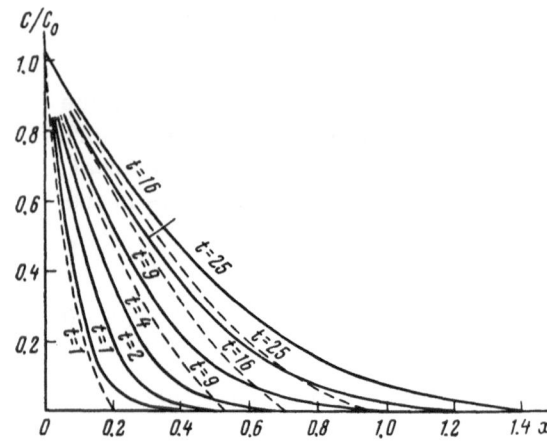

$$\frac{\partial C_1}{\partial t} = D_0 \frac{2n-1}{2n} \frac{\partial^2 C_1}{\partial x^2},$$

$$\frac{\partial C_2}{\partial t} = D_0 \frac{2n-3}{2n} \frac{\partial^2 C_2}{\partial x^2}, \qquad (6.15)$$

$$\cdots \cdots \cdots \cdots$$

$$\frac{\partial C_n}{\partial t} = D_0 \frac{2n-2i+1}{2n} \frac{\partial^2 C_n}{\partial x^2}.$$

By knowing the solution of Eq. (6.5) in the one-dimensional case for conditions (6.8) (Chap. 2):

$$C(x, t) = C_0 \left[1 - \operatorname{erf} \left(\frac{x}{2\sqrt{D_a t}} \right) \right], \qquad (6.16)$$

it is not difficult [5] to compute the concentration distribution of adsorbed substance along x, using Eq. (6.15). In Fig. 23 we have shown computed curves of concentration distributions of C and q along x for the linear (3.11) and convex (3.9) adsorption isotherms for different

Fig. 23. Curves of concentration distribution according to depth of grain (solid lines for linear adsorption isotherm, dashed lines for convex adsorption isotherm).

instants of time [5]. It is seen that the distribution curves of a substance for the linear and convex isotherms are appreciably different. The investigated "stepwise method" of solving the problem of diffusion in an adsorbing medium may be extended to any adsorption isotherm.

Similarly we may describe diffusion in nonporous microfractured rocks. In the body of these rocks occur microfractures filled with water. As a first approximation, we may assume that the microfractures are uniformly distributed in space. Diffusion is then described by the system (6.4) and (4.28) for definite initial and boundary conditions.

    2. Diffusion in a Discrete Porous Medium. Let us examine the diffusion of a solution or gas in a medium consisting of separate porous mineral grains capable of adsorbing the dissolved substance or of exchanging ions with it. This may be a decomposed rock, clay minerals, soils, and the like. The question of adsorption and ion-exchange properties of different minerals was considered in detail in Chap. 3.

We shall show how the problem is formulated in the simple case of one-dimensional diffusion of dissolved substance in a medium consisting of spherical mineral particles of radius r_0, with the intergranular spaces filled with water. At time $t = 0$ in the range $x > 0$, let there be no adsorbed substance, and, beginning at this instant, let a constant concentration of substance be maintained at the boundary $x = 0$ (diffusion from a steady source). The initial and boundary conditions of this problem have the form of Eq. (2.44). As follows from Chap. 4, the adsorption of dissolved substance by minerals may take place in the intergranular, intragranular, or mixed kinetic regions.

For describing diffusion migration it is convenient to use concentrations of substance in the bulk solution $C(x, t)$ and in the grain $q(x, t)$ averaged along the x profile. The equation of material balance (6.1) in this case assumes the form

$$\frac{\partial C}{\partial t} + \frac{\partial q}{\partial t} = \eta D \frac{\partial^2 C}{\partial x^2}. \qquad (6.17)$$

Let diffusion in the interspaces between grains take place more slowly than within the grains. The equation of intergranular kinetics of adsorption (ion exchange) may thus be found. In accordance with Fick's first law, it is easy to see that

$$\frac{\partial q}{\partial t} = \frac{4\pi r_0^2 D \left[\dfrac{\partial C\,(x,\,t,\,r)}{\partial r} \right]_{r=r_0}}{{}^4/_3 \pi r_0^3} = \frac{3}{r_0}\, D \left[\frac{\partial C\,(x,\,t,\,r)}{\partial r} \right]_{r=r_0}$$
$$(r \geqslant r_0),$$
(6.18)

where C (x, t, r) is the local concentration in the bulk solution at a distance r from the center of a grain.

The law of concentration distribution according to distance from the surface of a grain is found by solving the equation of nonstationary diffusion

$$\frac{\partial C\,(x,\,t,\,r)}{\partial t} = D \left[\frac{\partial^2 C\,(x,\,t,\,r)}{\partial r^2} + \frac{2}{r}\, \frac{\partial C\,(x,\,t,\,r)}{\partial r} \right] = \frac{D}{r}\, \frac{\partial^2\,[rC\,(x,\,t,\,r)]}{\partial r^2}$$
(6.19)

for the boundary conditions

$$C\,(x,\,t,\,r) = f\,(q), \quad r = r_0,$$
(on the surface of the grain)
$$C\,(x,\,t,\,r) = C\,(x,\,t), \quad r \to \infty,$$
(in the volume of the solution)
(6.20)

where C (x, t) is the average concentration in the x section.

The system (6.17)-(6.19) for conditions (2.44) and (6.20) describes one-dimensional diffusion geochemical migration in a discrete absorbent medium. The indicated system is complex. However, it may be simplified if we consider intergranular diffusion to be a quasistationary process (see Chap. 5). In this case

$$\left| \frac{\partial C\,(x,\,t,\,r)}{\partial t} \right| \ll \left| \frac{D}{r}\, \frac{\partial^2\,[rC\,(x,\,t,\,r)]}{\partial r^2} \right|$$
(6.21)

and in place of Eq. (6.19) we obtain

$$\frac{\partial^2\,[rC\,(x,\,t,\,r)]}{\partial r^2} = 0.$$
(6.22)

The solution of Eq. (6.22) for conditions (6.20) has the form

$$C\,(x,t,r) = C - [C - f\,(q)]\frac{r_0}{r}.$$
(6.23)

Substituting equation (6.23) in (6.18) we obtain the following equation of intergranular–diffusion adsorption kinetics:

$$\frac{\partial q}{\partial t} = \frac{3D}{r_0^2}\,[C - f\,(q)] = \gamma\,[C - f\,(q)].$$
(6.24)

The system (6.17) and (6.24) for conditions (2.44) approximately describes geochemical migration in a discrete porous medium in the intergranular region.

Let us investigate the system (6.17) and (6.24) for the linear adsorption isotherm f (q) = q/K. We shall exclude q, differentiating Eq. (6.17) according to time and substituting in Eq. (6.24):

$$\frac{\partial^2 C}{\partial t^2} + \gamma \left(1 + \frac{1}{K} \right) \frac{\partial C}{\partial t} = \eta D\, \frac{\partial^3 C}{\partial x^2 \partial t} + \frac{\eta D \gamma}{K}\, \frac{\partial^2 C}{\partial x^2}.$$
(6.25)

We shall consider adsorption (ion exchange) in minerals of low activity ($K \to 0$). Then the first terms on the left and right sides of Eq. (6.25) may be neglected. We shall have

$$\frac{\partial C}{\partial t} = \frac{\eta D}{1+K} \frac{\partial^2 C}{\partial x^2} = D_a \frac{\partial^2 C}{\partial x^2} \, , \tag{6.26}$$

which agrees with Eq. (6.5), written for the one-dimensional case. Thus, the solution of the problem of diffusion geochemical migration in a discrete porous medium as $K \to 0$ agrees with solutions of corresponding problems of diffusion migration in a continuous porous medium. A precise solution of the system (6.17) and (6.24) is very complicated [1].

Let us consider the case when diffusion within grains is the slower process. The equation of intragranular-diffusion adsorption (ion-exchange) kinetics is written for the linear isotherm of the system of equations

$$\frac{\partial q}{\partial t} = \frac{4\pi r_0^2 D_{ef} \left[\dfrac{\partial C\,(x,\,t,\,r)}{\partial r} \right]_{r=r_0}}{{}^4\!/_3 \pi r_0^3} = \frac{3}{r_0} D_{ef} \left[\frac{\partial C\,(x,\,t,\,r)}{\partial r} \right]_{r=r_0} \tag{6.27}$$
$$(r \leqslant r_0),$$

where $C\,(x,\,t,\,r)$ is the local concentration within the grain at a distance r from the center, found from equations of nonstationary diffusion:

$$\frac{\partial C\,(x,\,t,\,r)}{\partial t} = D_{ef} \left[\frac{\partial^2 C\,(x,\,t,\,r)}{\partial r^2} + \frac{2}{r} \frac{\partial C\,(x,\,t,\,r)}{\partial r} \right] = \frac{D_{ef}}{r} \frac{\partial^2 \left[rC\,(x,\,t,\,r) \right]}{\partial r^2} \tag{6.28}$$

for the boundary conditions

$$C\,(x,\,t,\,r) = KC\,(x,\,t), \quad r = r_0,$$
$$\frac{\partial C\,(x,\,t,\,r)}{\partial r} = 0, \quad r = 0 \tag{6.29}$$
$$\text{(condition of symmetry)}$$

$C\,(x,\,t)$ is, as before, the average concentration in a section of the medium.

The system (6.17) and (6.28) for conditions (2.44) and (6.29) describes geochemical migration of substance in the intragranular region. This system may be simplified if we consider intragranular diffusion to be a quasi-stationary process (Chap. 5). The equation of intragranular-diffusion adsorption (ion-exchange) kinetics may then be written approximately in the form (6.24) if we replace D by D_{ef}. Thus, as seen above, it may be shown that geochemical migration taking place in a discrete porous medium in the intragranular region is described asymptotically (when $K \to 0$) by the same equations as migration in a continuous porous medium.

3. Diffusion in a Discrete Medium, Considering Adsorption on the Surface. Let us consider the case of adsorption (ion exchange) taking place only on the surfaces of mineral grains. Glauconite (iron aluminum silicate containing potassium) is one of the substances free to exchange cations. It possesses a rigid crystal lattice with narrow cavities, and exchange therefore takes place only on the surface of the crystal (for more detail see Chap. 3).

In this case the process may take place only in the intergranular region of kinetics. If we consider the problem to be similar to that formulated in the previous subdivision of this section, it may then be described by a system of equations of material balance (6.17) and intergranular-diffusion kinetics of adsorption (ion exchange) for conditions (2.44). The kinetic equation of adsorption may be found in the following way. It is easy to see that

$$\frac{\partial q}{\partial t} = DS_V \left[\frac{\partial C(x,\,t,\,r)}{\partial r} \right]_{r=r_0},$$ (6.30)

where $C(x, t, r)$ is the local concentration in the bulk solution at a distance r from the center of the grain, and S_V is the area of a unit volume of the mineral.

The function $C(x, t, r)$ is found by solving Eq. (6.19) for conditions (6.20).

The system (6.17), (6.19), and (6.30) for conditions (2.44) and (6.20) describes one-dimensional diffusion geochemical migration, taking into account adsorption on the surface of mineral grains. The indicated system is complex. It may be simplified if we consider intergranular diffusion to be a quasi-stationary process. Then, by analogy with Eq. (6.24), we obtain the following equation of intergranular-diffusion adsorption (ion-exchange) kinetics:

$$\frac{\partial q}{\partial t} = DS_V r_0^{-1} [C - f(q)] = \gamma [C - f(q)].$$ (6.31)

§ 43. Diffusion in a Heterogeneous Medium, Considering Chemical Reactions

If the diffusing substance enters into chemical reactions with the rock, the process of interaction between substance and medium, as indicated in Chap. 4, may take place in the kinetic, diffusion, and mixed regions. The three types of problems that arise here, depending on the structural characteristics of the medium, were discussed in the preceding section. For brevity, we shall consider only diffusion in a continuous porous medium. The other types of problems may be described by a method similar to that discussed above for adsorption and ion exchange.

Let the rate of chemical reaction be less than the diffusion rate; i.e., the process takes place in the kinetic range. Geochemical migration of substance in this case is described by the system of equations (6.1) and the kinetic equations of corresponding chemical reaction for definite initial and boundary conditions.

If the reaction between the diffusing substance and the rock is first-order irreversible, then, in accordance with Eq. (4.7), the equation of reaction rate may be written in the form

$$\frac{\partial q}{\partial t} = KC.$$ (6.32)

The solution of the system (6.1) and (6.32) for one-dimensional diffusion along the x axis from a steady source [condition (6.8)] has the form [7]:

$$C(x, t) = C_0 \left\{ e^{-Kt} \left[1 - \text{erf} \left(\frac{x}{\sqrt{2Dt}} \right) \right] + Kt \int_0^1 e^{-Kty} \left[1 - \text{erf} \left(\frac{x}{\sqrt{2Dty^{1/2}}} \right) \right] dy \right\}.$$ (6.33)

The solution of the system of equations (6.1) and (6.32) for other initial and boundary conditions has been examined by other authors [1, 8].

If a first-order reversible reaction takes place between the diffusing substance and the rock, the rate equation, in accordance with Eq. (4.13), is written in the following form:

$$\frac{\partial q}{\partial t} = K_1 C - K_2 q = K_1 \left(C - \frac{q}{K_c} \right),$$ (6.34)

where $K_c = K_1/K_2$ is the equilibrium constant of the investigated reaction.

Equation (6.34) is analogous to (6.24), the equation of intergranular-diffusion adsorption kinetics for the linear isotherm, differing only in the value of the factor in front of the concentration gradient $(C - q/K_c)$. Hence, it follows that if K_c is small, geochemical migration of the substance is described in the one-dimensional case of Eq. (6.26). Where we understand K to mean the equilibrium constant of the reaction (5.46), K_c. The solutions of Eq. (6.26) have the forms of (2.33), (2.40), (2.45), (2.47), and (2.54), depending on the investigated problem, if we replace D by the value $D/(1 + K_1/K_2)$.

The solution of more complex diffusion problems with consideration of chemical reactions have been discussed in other papers [1, 8-12].

§44. Filtration in a Heterogeneous Medium in the Absence of Interaction between Substance and Medium

Let a substance in the liquid or gaseous phase move through rocks at some rate $\vec{u}$ by virtue of a pressure drop. If the substance reacts weakly with the rocks and is insignificantly adsorbed by them $(K \to 0)$, the process may be considered as if there were approximately no interaction between the substance and the medium.

For simplicity we shall restrict our examination to the one-dimensional problem, in which the flow of substance is along the x axis at constant rate u. In this case the concentration of substance at all points in a plane perpendicular to x will be the the same, so that $\partial C/\partial y = \partial C/\partial z = 0$. Since exchange of material between mobile and fixed phases is absent, then, in place of Eqs. (1.9) and (1.17) we shall have

$$\frac{\partial C}{\partial t} + u\,\frac{\partial C}{\partial x} - \eta D\,\frac{\partial^2 C}{\partial x^2} = 0. \tag{6.35}$$

Equation (6.35) of convective diffusion describes for certain initial and boundary conditions the distribution of dissolved substance flowing along the x axis in a porous medium. If filtration of the substance takes place through a porous granular medium, the process is affected by the disorder in grain distribution and the inhomogeneity of the pores, leading to supplementary spreading of the front of the migrating solution.

The inhomogeneity of the medium is a statistical factor. Therefore, the displacement of molecules of dissolved substance because of any of the indicated factors is equally probable in any direction. As we shall show in Chap. 7 (on the basis of the formation of a mechanical dissemination aureole at an ore deposit), the results of the effect of the probability factors are equivalent to the action of diffusion. Consequently, the inhomogeneity of a medium may be taken into account by introducing in (6.35) the coefficient of longitudinal diffusion D^* in place of ηD, considering all the spreading factors.

Let us consider the solution to the following problems of the filtration of solutions through rocks: 1) filtration from a semiinfinite bed, 2) filtration from a steady source, and 3) filtration from an infinitesimally thin layer in an unbounded body.

1. Filtration from a Semiinfinite Bed. Let the initial concentration distribution in the medium be such that for all abscissa values less than zero $C(x, 0) = C_0 = const$, and for all abscissas greater than zero $C(x, 0) = 0$. Physically, this means filtration from a very large (semiinfinite) bed possessing porosity or fracturing like the surrounding rocks. The solution of Eq. (6.35) for such conditions may be written in the form [13, 14]

$$C\left(x,t\right)=\frac{C_0}{2}\left[1-\mathrm{erf}\left(\frac{x-ut}{2\sqrt{D^*t}}\right)\right],\tag{6.36}$$

where $\mathrm{erf}\left(\frac{x-ut}{2\sqrt{D^*t}}\right)$ is the Gaussian integral (2.34).

From Eq. (6.36) it follows that during filtration a front of substance that grows diffuse with time is formed; it is shown schematically in Fig. 24. The distance at which the concentration of dissolved substance changes from 0 to C_0 increases with time.

For geochemistry, it is interesting to consider the distance penetrated by the dissolved substance from the bed into the surrounding rock. We may write without any great error (see Chap. 2)

$$C\left(x,\ t\right)=\frac{C_0}{2}\left[1-\left(\frac{x-ut}{\sqrt{\pi D^*t}}\right)\right],\tag{6.37}$$

whence for minimal concentration, which may still be determined by quantitative analysis,

$$C_{\min}=\frac{C_0}{2}\left[1-\left(\frac{x_{\max}-ut}{\sqrt{\pi D^*t}}\right)\right],\tag{6.38}$$

where $x_{\max}$ is the distance of migration (see Chap. 2).

Hence

$$x_{\max}=ut+\left(1-\frac{2C_{\min}}{C_0}\right)\sqrt{\pi D^*t}.\tag{6.39}$$

If $C_{\min}\ll C_0$, then

$$x_{\max}=ut+\sqrt{\pi D^*t}.\tag{6.40}$$

From Eqs. (2.37) and (6.40) it follows that during filtration the distance of migration is greater than that by diffusion penetration by the value ut. By using Eq. (6.40) it is possible to calculate the maximum distance a substance will move by filtration in a given time.

Let us consider the conclusions deriving from Eq. (6.36) [13].

1. The equation of movement of points of constant concentration at the front of a salt

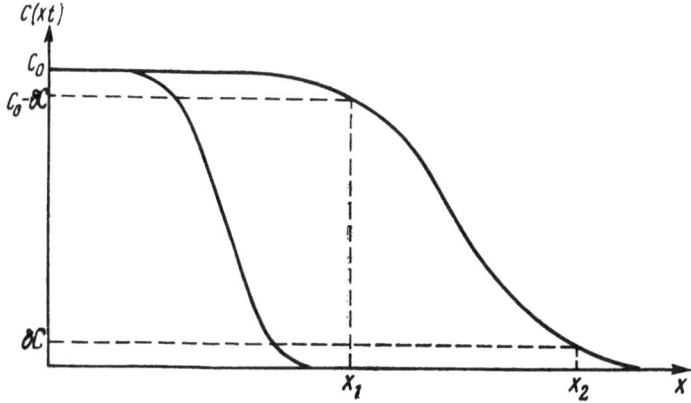

Fig. 24. Distribution of substance during filtration
from a semiinfinite bed.

solution may be written in the form

$$x_c = ut + 2z \sqrt{D^*t}, \tag{6.41}$$

where $z = z(C/C_0)$ is the argument of the Gaussian integral (2.34), depending on the relative concentration C/C_0.

2. When

$$\frac{C}{C_0} = 0.5, \quad \mathrm{erf}\left(\frac{x - ut}{2\sqrt{D^*t}}\right) = 0,$$

hence

$$x_{1/2 C_0} = ut. \tag{6.42}$$

Consequently, among the concentration points, only one, the point of half concentration, $C = 0.5\,C_0$, moves at a steady rate equal to u. The other points in the front move at variable rates:

$$\left(\frac{dx}{dt}\right)_C = u + z \sqrt{\frac{D^*}{t}}, \tag{6.43}$$

$$\text{while} \quad \left(\frac{\partial x}{\partial t}\right)_C > u \quad \text{for} \quad \frac{C}{C_0} < 0.5;$$

$$\left(\frac{\partial x}{\partial t}\right)_C < u \quad \text{when} \quad \frac{C}{C_0} > 0.5.$$

This latter means that the front of dissolved substance is continuously spread. It is more diffuse the greater the coefficient of longitudinal diffusion D^*.

3. Let us consider how the length of the front depends on time. By length of the front λ we mean the distance at which the concentration in the front changes from δC to $C_0 - \delta C$, where δC is the least concentration that may be determined by analytical methods. In Fig. 24, $\lambda = x_2 - x_1$, where $C(x_2) = \delta C$, $C(x_1) = C_0 - \delta C$. From the properties of the Gaussian integral (2.34), it follows that $z = z(\delta C/C_0) = -z(1 - \delta C/C_0)$. Then, from Eq. (6.41) we obtain the following equation for the length of the front:

$$\lambda = \frac{2}{x_1} - \frac{1}{x_2} = 4z(\delta C) \sqrt{D^*t}, \tag{6.44}$$

where the independent variable z is taken at $C = \delta C$.

From the indicated equation it follows that the length of the front of dissolved substance increases proportionally to $\sqrt{t}$.

Let us examine the solution of Eq. (6.35) of convective diffusion for the problem of elution by water of the dissolved substance filling the free space of the rocks in the zone $x \geq 0$. Let the initial concentration distribution be given in the form

$$t = 0, C(x, 0) = \begin{cases} C_0, & x \geq 0, \\ 0, & x < 0. \end{cases} \tag{6.45}$$

The solution of Eq. (6.35) for conditions (6.45) is written in the following form:

$$C(x, t) = \frac{C_0}{2}\left[1 + \mathrm{erf}\,\frac{x - ut}{2\sqrt{D^*t}}\right]. \tag{6.46}$$

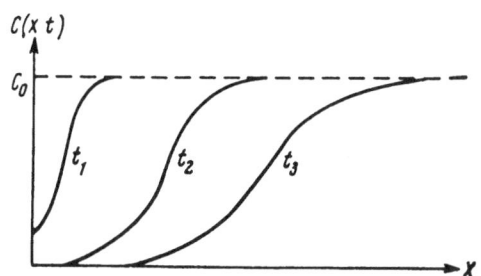

Fig. 25. Distribution of substance during elution of a semiinfinite bed.

Equation (6.46) is similar to (6.36). They differ only in the sign of the independent variable of the Gaussian integral (2.34) (since $-\operatorname{erf} z = \operatorname{erf}(-z)$). This means that the curve of concentration distribution during elution of a semiinfinite bed (Fig. 25) increases with increase in x (in contrast to Fig. 24). Therefore, for a theoretical description of the elution of dissolved substance from a semiinfinite bed we may use Eqs. (6.41)-(6.44) if we change the sign of the independent variable to the opposite value.

2. Filtration from a Steady Source.
We shall formulate the problem of filtration of dissolved substance from a steady source. In the section at x = 0 let us place the boundary between rocks and a large bed, and at this boundary we shall maintain a steady concentration of dissolved substance (for example, equal to the concentration of a saturated solution). The initial and boundary conditions of the problem have the form

$$x = 0, \quad t > 0, \quad C = C_0,$$
$$t = 0, \quad x > 0, \quad C = 0. \tag{6.47}$$

The solution of Eq. (6.35) for conditions (6.47) has the form [14]

$$C(x, t) = \frac{C_0}{2}\left\{\left[1 - \operatorname{erf}\left(\frac{x - ut}{2\sqrt{D^*t}}\right)\right] + e^{ux/D}\left[1 - \operatorname{erf}\left(\frac{x + ut}{2\sqrt{D^*t}}\right)\right]\right\}. \tag{6.48}$$

If x and t are rather large, then, expanding $\operatorname{erf}\left(\frac{x + ut}{2\sqrt{D^*t}}\right)$ into a series, it may be shown [14] that the second term in Eq. (6.48) is small as compared with the first and may be neglected. Therefore the asymptotic solution (for large values of x and t) of Eq. (6.35) for conditions (6.47) has the form (6.36). The distribution of dissolved substance during filtration from a steady source asymptotically obeys the same rules [Eqs. (6.39)-(6.44)] as filtration from a semiinfinite bed. Equation (6.48) also describes the filtration of gas from a large bed.

3. Filtration from an Infinitesimally Thin Layer in an Unbounded Body. Let us examine the solution of Eq. (6.35) for the following initial and boundary conditions. Let a solvent flow continuously along the x axis, and, at the moment t = 0, let a small (infinitesimally small at the extreme) of substance with concentration C_0 (filtration from an instantaneous source or an infinitesimally thin layer) be "injected" into the current. The concentration distribution in the current at any instant will be described by the following function [14, 15]:

$$C(x, t) = \frac{C_0}{2\sqrt{\pi D^* t}}\, e^{\frac{-(x - ut)^2}{4D^*t}}. \tag{6.49}$$

From Eq. (6.49) it follows that the band of dissolved substance introduced into the current is continuously spread with time. A maximum concentration appears in the band, moving at constant rate u and decreasing in value with time.

The problem of one-dimensional filtration of solutions has also been solved for those cases in which the concentration of the substance in the current entering the investigated zone is distributed according to the law of straight and broken lines [16-18].

In the case of multidimensional filtration the problem is substantially more complex. From qualitative considerations it follows that diffuseness of the front of dissolved substance will be determined here not only by diffusion, but also by the shape of the filter and the method of supply and removal of solution relative to the filter. Diffuseness because of these factors is due to the fact that different elements in the filtering bulk solution pass through the filter along unlike paths because of variations in method of supply and withdrawal of solution and differences in filter shape. Therefore, different volumes of filtering solution reach the outlet of the filter at any time, as a result of which solutions of different concentrations are mixed there. This leads to supplementary spreading of the front of dissolved substance.

§ 45. Filtration of a One-Component Solution in a Porous Medium, Considering Adsorption and Ion Exchanges

Let us examine the more general problem of movement of a one-component solution (liquid or gaseous) through a porous medium in the presence of adsorption and ion-exchange interaction between the substance and the medium. In its general form this problem was formulated in Chap. 1 [the system (1.9) and (1.17)]. Below we shall consider analytical solutions of the problem, which have been obtained only for the one-dimensional case. The evolved theory describes the filtration of solutions and gases under natural conditions, since rocks surely possess adsorption capacity (Chap. 3).

Let a substance move through a rock in a direction along the x axis at a steady rate u. We are required to find the law of concentration distribution of substance in the medium at any instant of time when we know the initial distribution (at t = 0). The solution of the problem we have raised may be found by using the equation of material balance of a moving substance and the kinetic equation of adsorption (ion exchange) in a current. Numerous investigations (experimental) on adsorption and ion exchange have shown that the rates of these processes are generally determined by diffusion (Chap. 5). Consequently, the investigated process may take place in the intergranular, intragranular, and mixed kinetic regions.

Let us find a solution of the problem for filtration in a medium systematically built of spherical mineral grains as shown in Fig. 26, and point out how to extend the theory to more complex cases. A similar problem is solved in the theory of chromatography [19-22], and is called the problem of the dynamics of adsorption.

Let us take a unit volume V for which we write the equation of material balance and the kinetics of adsorption (ion exchange); the unit volume is of a rock layer one grain thick and perpendicular to the x axis (Fig. 26). The equations of diffusion kinetics of adsorption (ion

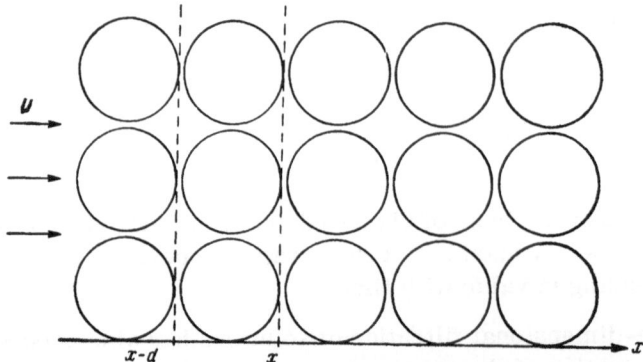

Fig. 26. Schematic illustration of medium of systematically arranged spherical mineral grains.

exchange) from a current in a layer with coordinates $(x - d, x)$, where d is the diameter of a grain, was obtained in Chap. 5 [Eqs. (5.28) and (5.27)]. The equation of material balance for this layer is written in the form [21, 23]

$$\frac{\partial C(x,\, t)}{\partial t} + \frac{u}{d}\left[C(x,\, t) - C(x - d,\, t)\right] + \frac{\partial q(x,\, t)}{\partial t} = \frac{\eta D}{d}\left[-\frac{\partial C(x,\, t)}{\partial x} + \frac{\partial C(x - d,\, t)}{\partial x}\right]. \qquad (6.50)$$

Expanding $C(x - d, t)$ and $\partial C(x - d, t)/\partial x$ into series for powers of d and restricting ourselves to derivatives of the second order, we obtain in place of Eq. (6.50) the following approximate equation:

$$\frac{\partial q}{\partial t} + u\frac{\partial C}{\partial x} + \frac{\partial C}{\partial t} = \left(\eta D + \frac{ud}{2}\right)\frac{\partial^2 C}{\partial x^2}. \qquad (6.51)$$

The system (5.27), (5.28), and (6.51) describes for certain initial and boundary conditions the filtration of dissolved substance in an adsorbing medium. The solution of this system depends on the form of the adsorption (ion-exchange) isotherm.

Tentatively we may distinguish three types of isotherms (see Fig. 9): convex (1), linear (2), and concave (3). If the concentration of filtering substance is small (as ordinarily observed in nature), the adsorption isotherm is linear. The ion-exchange isotherms for rocks ordinarily differ little from linearity (Chap. 3). Therefore, the solution of the system (5.27) and (6.51) for a linear isotherm (3.11) is of special interest.

Let us find a solution of the indicated system for a linear adsorption (ion-exchange) isotherm and for different initial and boundary conditions on the basis of the concept of effective longitudinal diffusion [22, 23]. According to this concept, the diffuseness of the front of filtering substance because of intergranular and intragranular diffusion may be described as effective longitudinal diffusion of a substance with a diffusion coefficient D_{lo}. For this we shall show that the system of equations (5.27) and (6.51) is equivalent to

$$\frac{\partial q}{\partial t} + u\frac{\partial C}{\partial x} + \frac{\partial C}{\partial t} = (\eta D + ur_0 + \lambda)\frac{\partial^2 C}{\partial x^2} = D_{lo}\frac{\partial^2 C}{\partial x^2}, \qquad (6.52)$$

$$\frac{\partial q}{\partial t} = K\frac{\partial C}{\partial t}, \qquad (6.53)$$

where λ is a value taking into account the diffuseness of the front because of intergranular and intragranular diffusion.

Differentiating Eq. (5.27) according to t and substituting the expression from Eq. (6.51) in place of $\partial q/\partial t$, we obtain

$$\frac{\partial^2 q}{\partial t^2} = -\frac{\gamma}{K}\left[(1 + K)\frac{\partial C}{\partial t} + u\frac{\partial C}{\partial x} - (\eta D + ur_0)\frac{\partial^2 C}{\partial x^2}\right]. \qquad (6.54)$$

In order that Eq. (6.54) agree with Eqs. (6.52) and (6.53), it is necessary that

$$\frac{\partial^2 q}{\partial t^2} = \frac{\lambda\gamma}{K}\frac{\partial^2 C}{\partial x^2}. \qquad (6.55)$$

We shall transform the right side of Eq. (6.55) to the following form:

$$\frac{\lambda\gamma}{K}\frac{\partial^2 C}{\partial x^2} = \frac{\lambda\gamma}{K}\left(\frac{\partial^2 C}{\partial t^2}\right)\left(\frac{dt}{dx}\right)_C^2 = \frac{\lambda\gamma}{K^2}\left(\frac{\partial^2 q}{\partial t^2}\right)\left(\frac{dt}{dx}\right)_C^2. \qquad (6.56)$$

We shall seek a solution to the system of equations (5.27) and (6.51) for filtration from a steady source (infinite bed). Let us place the bed so that its boundary with the surrounding rocks is in the plane x = 0, and the x axis, along which filtration occurs, is perpendicular to the bed. We shall assume that at t = 0 in the region x > 0 the migrating substance (of the bed) is absent. A constant concentration C_0 of substance of the bed is maintained at the boundary x = 0, beginning at t = 0. Mathematically these conditions may be written in the following form:

$$t=0 \begin{cases} x>0, \quad C=0, \quad q=0, \\ x<0, \quad C=C_0, \end{cases}$$
$$x=0, \quad t>0, \quad C=C_0. \tag{6.57}$$

The corresponding problem in the theory of chromatography is called frontal analysis. Let us consider that the asymptotic solution of the system (6.52) and (6.53) for conditions (6.57) may be borrowed from the solution of the problem of convective diffusion [14]. It has the form

$$C(x,\,t)=\frac{C_0}{2}\left[1-\mathrm{erf}\left(\frac{x-\frac{u}{1+K}t}{2\sqrt{\frac{D_{\mathrm{lo}}}{1+K}t}}\right)\right]. \tag{6.58}$$

From the derived equation it follows that the point of half concentration in the front, $C = 0.5C_0$, moves at a constant rate equal to $u/(1 + K)$. For points near the point of half concentration, then, the following approximate equality is fulfilled:

$$\left(\frac{dx}{dt}\right)_C \approx \frac{u}{1+K}. \tag{6.59}$$

Substituting the value $\partial x/\partial t$ from Eq. (6.59) in Eqs. (6.56) and (6.55), we obtain the following expression for λ:

$$\lambda \approx \frac{K^2}{(1+K)^2}\,\frac{u^2}{\gamma}. \tag{6.60}$$

From the above discussion it follows that the asymptotic solution of the problem of one-dimensional filtration from a steady source or from an infinite bed [conditions (6.57)] may be written in the form (6.58), in which the effective coefficient of longitudinal diffusion, in accordance with Eqs. (6.52), (5.27), and (5.28), is equal to

$$D_{\mathrm{lo}} = \eta D + u r_0 + \frac{K^2}{(1+K)^2}\,0.53 D^{-1/3} u^{5/3} r_0^{5/3} + \frac{K}{(1+K)^2}\,\frac{1}{12}\,D_{\mathrm{ef}}^{-1} u^2 r_0^2. \tag{6.61}$$

Figure 27 illustrates the concentration distribution of adsorbed substance in the medium for different times on the basis of Eq. (6.58). For comparison, the dashed lines on this graph represent the distribution of dissolved substance in the absence of interaction with the rocks. It is seen that the front has become diffuse, and in it the point of half concentration moves at a constant rate less than the flow rate:

$$v = \frac{u}{1+K}, \tag{6.62}$$

the remaining points move with variable velocities. The diffuseness of the front is greater the higher the effective coefficient of longitudinal diffusion D_{lo}. In accordance with Eq. (6.61),

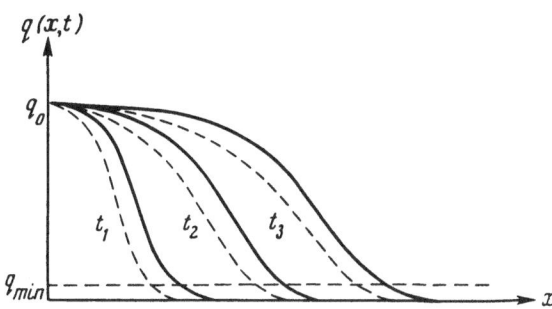

Fig. 27. Distribution of adsorbed substance during filtration from a steady source.

the diffuseness of the front during one-dimensional filtration is determined by the following factors: 1) longitudinal diffusion, 2) size of grains in the rock, 3) intergranular diffusion, and 4) intragranular diffusion.

The illustration in Fig. 27 should be expanded for the ion-exchange picture. During ion exchange, one kind of ion (from the solution) is adsorbed, the other kind (from the rock) is set free. The picture in Fig. 27 gives the distribution for the ion initially found in the solution. By virtue of the preservation of electroneutrality during ion exchange, in any infinitesimally small volume of the medium the following condition must be fulfilled:

$$\begin{cases} C_1(x,\ t) + C_2(x,\ t) = C_0 \\ q_1(x,\ t) + q_2(x,\ t) = q_0, \end{cases} \tag{6.63}$$

where the subscript 1 refers to one kind of ion, 2 to the other.

Consequently, the distribution of ions displaced from the rock, in accordance with (6.58), is written in the form

$$C(x,\ t) = \frac{C_0}{2}\left[1 \div \mathrm{erf}\left(\frac{x - \dfrac{u}{1+K}t}{2\sqrt{\dfrac{D_{1o}}{1+K}t}}\right)\right]. \tag{6.64}$$

The distribution of ions in accordance with Eq. (6.64) is similarly illustrated in Fig. 25.

Let us examine the case in which, in expression (6.61), all terms are small in comparison with ur_0 and may be neglected. It is easy to see that this obtains only for adsorbents of low activity and when flow rates are not too low, i.e., for the conditions

$$K \to 0$$
$$u \gg \frac{D}{r_0}. \tag{6.65}$$

In this case the rates of intergranular and intragranular diffusion are large, and at each point in the medium adsorption equilibrium is established at any instant of time. Since longitudinal diffusion is slight, the supply rate of substance to the surface of the grains is equal to the flow rate. In this situation adsorption takes place at the proper time at every point in the medium: the rate of accumulation of adsorbed substance is determined by the flow rate. We may show that the equation of material balance (6.50) for conditions (6.65) changes to the kinetic equation (5.32) because of flow. Substituting from Eq. (6.53) in (6.50) and taking into account that longitudinal diffusion is slight, we obtain

$$\frac{\partial q}{\partial t} = \frac{K}{1+K}\frac{u}{d}[C(x-d,\ t) - C(x,\ t)]. \tag{6.66}$$

Since equilibrium obtains in the layer (x− d, x), in place of Eq. (6.66) we may write

$$\frac{\partial q}{\partial t} = \frac{K}{1+K}\frac{u}{d}\left[C - \frac{q}{K}\right], \tag{6.67}$$

where C is the concentration of substance feeding into layer (x− d, x).

Equation (6.67) is in agreement with (5.32).

In generalizing the results we have obtained, it may be stated that Eq. (6.58) describes the filtration of a substance in any medium (continuous or discrete porosity) in which the effective coefficient of longitudinal diffusion is equal to

$$D_{lo} = D^* + \frac{K^2 u^2}{(1+K)^2}\left(\frac{1}{\gamma_1} + \frac{1}{\gamma_2}\right), \tag{6.68}$$

where γ_1 and γ_2 are the rate constants of intergranular and intragranular diffusion respectively.

The coefficient of longitudinal diffusion D^* depends on the coefficient of diffusion D in the bulk solution, the inhomogeneity of the medium, and the grain size of the rock. The distance of migration of dissolved substance is found in the same way as in Eq. (6.39):

$$x_{\max} = \frac{u}{1+K}t + \left(1 - \frac{2C_{\min}}{C_0}\right)\sqrt{\pi\frac{D_{lo}}{1+K}t}. \tag{6.69}$$

If $C_{\min} = \alpha C_0$ then

$$x_{\max} = \frac{u}{1+K}t + \sqrt{\pi\frac{D_{lo}}{1+K}t}\,(1-2\alpha). \tag{6.70}$$

A comparison of Eqs. (6.70) and (6.40) shows that adsorption diminishes the distance the dissolved substance will migrate. Similarly, an expression is found for the distance of migration in the solid phase if $C_{\min}$ and C_0 in Eqs. (6.69) and (6.70) are replaced by $C_{\min}$ and q_0.

Let us consider the solution of the system (5.27) and (6.51) for other initial and boundary conditions. Let a small amount of dissolved substance (infinitesimally small at the limit) be introduced into the current at time t = 0. We shall write the initial and boundary conditions of the problem in the form

$$t = 0, \quad \begin{cases} 0 \leqslant x \leqslant x_0, & C = C_0, \quad q = KC_0, \\ x > x_0, & C = 0, \quad q = 0, \end{cases}$$
$$x = 0, \quad t > 0, \quad C = 0, \quad q = 0. \tag{6.71}$$

The system (5.27) and (6.51) for the indicated conditions describes the filtration of substance from an instantaneous source or from an infinitesimally thin layer. The corresponding problem in chromatography is called elution analysis. The solution of this problem may be found similar to that of filtration from a steady source, and will have the form

$$C(x, t) = \frac{C_0}{2\sqrt{\pi\frac{D_{lo}}{1+K}t}}\exp -\left(\frac{x - \frac{u}{1+K}t}{2\sqrt{\frac{D_{lo}}{1+K}t}}\right)^2, \tag{6.72}$$

where the effective coefficient of longitudinal diffusion is determined by Eq. (6.61) or (6.68) depending on the geometrical parameters of the medium. From Eq. (6.72) it follows that the band of adsorbed substance during movement is continuously spread. A maximum concentration appears in the band, which moves at a constant rate u/(1 + K), the value of which decreases with time (see below).

The solution of the problem of one-dimensional filtration for conditions (6.57) and (6.71) in the case of a nonlinear adsorption isotherm may be borrowed from the corresponding solutions of the problem of the dynamics of adsorption and chromatography [24-30]. It should be noted that we have obtained asymptotic solutions (as $t \to \infty$) only for the convex adsorption (ion-exchange) isotherm (see Fig. 9). Analytical solutions for the concave isotherm are lacking. Without dwelling on the solutions [24-30], let us consider the qualitative results given by the theories of adsorption dynamics for a nonlinear isotherm.

For conditions (6.57) a stationary front of dissolved substance (not changing form in the course of time) has been established in the case of the convex adsorption (ion-exchange) isotherm; the front moves at constant rate:

$$V = \frac{q_0}{C_0 + q_0}\, u, \tag{6.73}$$

where q_0 is the adsorption capacity of the rock.

With a concave adsorption (ion-exchange) isotherm, a continuously diffuse front of filtering substance is formed.

§ 46. Geochemical Migration due to Filtration in a Weakly Adsorbent Medium

We shall examine the case when solutions circulating through the rocks are found in ore deposits or gas deposits and are therefore rather highly concentrated. Figure 14 shows a typical adsorption isotherm of gases and liquids of the Langmuir type for porous minerals. As seen from the figure, the amount of adsorbed substance q (per unit volume of mineral) reaches a limiting value $q = q_0$ at a solution concentration of $C = C_s$. Below we shall consider the case in which the concentration of filtering solution is considerably greater than C_s ($C^* \gg C_s$), and the surrounding rocks will have a low adsorption capacity (q_0 is low). In this case, the adsorption isotherm may be represented approximately by a straight line (see Fig. 15).

Formulation of the Problem. Let a substance in the liquid or gaseous phase move through rocks at some rate u because of a drop in pressure. We shall assume that at time t = 0 we know the distribution of substance in the medium. As a consequence of filtration and adsorption, the distribution changes with time. The problem of geochemical migration consists in determining the distribution function in the medium at any instant of time. As in the previous problem, this is solved by using the equation of material balance and the kinetic equation of adsorption in a current.

For simplicity we shall consider the one-dimensional problem, in which the flow of substance is along the x axis at a constant rate u. We shall assume that the surrounding rock consists of identical spherical grains of radius r_0, uniformly arranged in space. As a unit layer, for which we shall write equations of material balance and of adsorption kinetics, it is expedient to select a layer of the surrounding rock one grain thick, perpendicular to the x axis (see above). The equation of material balance for this layer with the coordinates (x−d, x) has the form (6.51).

We shall derive the kinetic equation of adsorption for this case. Since the adsorption isotherm has the form illustrated in Fig. 15, at any value C > 0 the concentration of adsorbed substance on the grain surfaces of the rock is equal to q_0. Because of the low adsorption capacity of the rock, the concentration of adsorbed substance on the grain surfaces differs little from the concentration in the bulk solution. It may be assumed, approximately, that there is no concentration gradient in the bulk solution in a direction perpendicular to the surface of a grain, and, consequently, the intergranular-diffusion stage is absent in the adsorption kinetics. The process in this case takes place in the intragranular-diffusion region (see Chap. 5).

Let us assume a linear concentration distribution of the adsorbed substance through the thickness of a grain at any instant of time [31]. Then, for the time $t \geq \tau$, where τ is the time required for the concentration front to reach the center of the grain, we have, in accordance with Fick's first law,

$$\frac{\partial Q}{\partial t} = 4\pi r_0^2 D_{ef} \frac{q_0 - q(0)}{r_0} = 4\pi r_0 D_{ef} [q_0 - q(0)], \qquad (6.74)$$

where Q is the amount of adsorbed substance adsorbed by the grain at time t, q(0) is the concentration of substance adsorbed by the grain at the center of the grain (r = 0), and D_{ef} is the effective coefficient of intergranular diffusion. With a linear concentration distribution in the function of r, the relationship between Q and q(0) is given by Eq. (5.17).

Substituting from Eq. (5.17) in equation (6.74) with $Q = \frac{4}{3}\pi r_0^3 q$, we obtain the following equation of intragranular-diffusion adsorption kinetics [comparable with (5.19)]:

$$\frac{\partial q}{\partial t} = \frac{12 D_{ef}}{r_0^2}(q_0 - q). \qquad (6.75)$$

Investigation has shown that Eq. (6.75) may be used to describe adsorption also at $t < \tau$.

The system (6.51) and (6.75) describes geochemical migration, for certain initial and boundary conditions, when the migration is due to filtration in a weakly adsorbent medium made up of identical mineral grains. We may similarly formulate the problem for the more general case in which the "ideal character" of the medium (identical mineral grains, uniform distribution of grains in space) is lacking. The process is described by the equation of material balance of a moving substance:

$$\frac{\partial q}{\partial t} + u \frac{\partial C}{\partial x} + \frac{\partial C}{\partial t} = D^* \frac{\partial^2 C}{\partial x^2} \qquad (6.76)$$

and the equation of intragranular-diffusion adsorption kinetics:

$$\frac{\partial q}{\partial t} = \gamma [q_0 - q], \qquad (6.77)$$

where γ is the rate constant of intergranular diffusion, depending on the coefficient of intragranular diffusion and the geometrical shapes and sizes of the mineral grains, and D^* is the coefficient of longitudinal diffusion, taking into account the spreading (nonsharpening) factors at work in the mobile phase.

It is seen that the system of equations (6.51) and (6.75) is a particular case of the more general system (6.76) and (6.77).

We shall seek a solution of system (6.76) and (6.77) for filtration from a steady source [conditions (6.57)].

Solution of the Problem without Consideration of Longitudinal Diffusion. Let D^* be small; then the number $D^* \partial^2 C/\partial x^2$ in Eq. (6.76) may be neglected. We shall substitute in Eqs. (6.76) and (6.77) the independent variables

$$\begin{cases} x' = x, \\ t' = t - \dfrac{x}{u}. \end{cases} \qquad (6.78)$$

Then Eqs. (6.76), (6.77), and (6.57), without taking longitudinal diffusion into account, acquire the form

$$u\,\frac{\partial C}{\partial x'} + \frac{\partial q}{\partial t'} = 0,$$ (6.79)

$$\frac{\partial q}{\partial t'} = \gamma\,(q_0 - q),$$ (6.80)

$$C\,(0,\,t') = C_0,\quad t' \geq 0,$$ (6.81)

$$q\,(x',\,0) = 0,\quad x' \geq 0.$$ (6.82)

The solution of Eq. (6.80), taking Eq. (6.82) into account, is written in the following form, with a change to the variables x and t:

$$q\,(x,\,t) = \begin{cases} q_0\left[1 - e^{-\gamma\left(t - \frac{x}{u}\right)}\right], & t > \frac{x}{u}, \\ 0, & t < \frac{x}{u}. \end{cases}$$ (6.83)

Differentiating Eq. (6.83) according to t, substituting in Eq. (6.79), and integrating the resulting equation for conditions (6.81), we obtain

$$C\,(x,\,t) = \begin{cases} C_0 - \frac{q_0 \gamma}{u}\,x e^{-\gamma\left(t - \frac{x}{u}\right)}, & t \geq \frac{x}{u}, \\ 0, & t < \frac{x}{u}. \end{cases}$$ (6.84)

Equations (6.83) and (6.84) give a solution of the problem without consideration of effective longitudinal diffusion. In Figs. 28 and 29 we have shown the distribution curves for the substance in the rock and in the mobile phase along the line x at different times. From Eq. (6.84) and Fig. 29 it follows that a diffuse front of adsorbed material is formed. The foremost point in the front moves at the rate u.

Solution of the Problem with Consideration of Longitudinal Diffusion. Let us consider the case in which in Eq. (6.76) the member $D^* \partial^2 C / \partial x^2$ cannot be

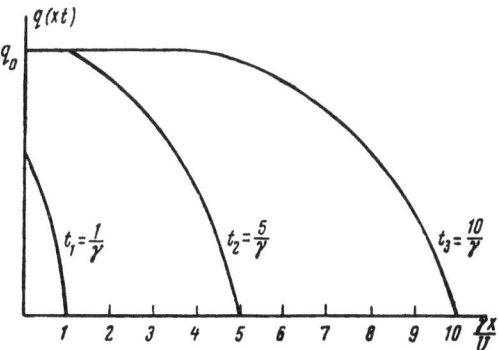

Fig. 28. Distribution curves of a substance in the immobile phase during filtration in a weakly adsorbent medium.

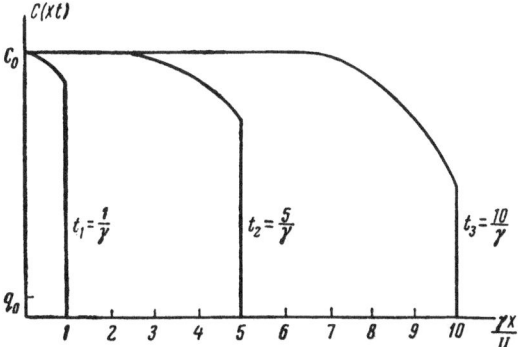

Fig. 29. Distribution curves of a substance in the mobile phase during filtration in a weakly adsorbent medium.

thrown out. But, since the adsorption capacity of the adsorbent is small and the concentration of the solution is large, the derivative $\partial C/\partial t$ in this equation is small as compared with $\partial q/\partial t$, and it may be neglected. The solution for $q(x, t)$ is, as before, Eq. (6.83). Substituting in Eq. (6.76) the value from (6.83) in place of $\partial q/\partial t$, and rejecting $\partial C/\partial t$, we have for $t > x/u$

$$u \frac{\partial C}{\partial x} + D^* \frac{\partial^2 C}{\partial x^2} = - q_0 \gamma e^{-\gamma t}. \qquad (6.85)$$

Integrating Eq. (6.85), we obtain

$$C - \frac{D^*}{u} \frac{\partial C}{\partial x} = - \frac{q_0 \gamma}{u} x\, e^{-\gamma t} + \varphi(t), \qquad (6.86)$$

where $\varphi(t)$ is some function of t.

Since at $x = 0$, $C = C_0$, $\partial C/\partial x \approx 0$, then $\varphi(t) \approx C_0$ and in place of (6.86) we have

$$\frac{\partial C}{\partial x} - \frac{u}{D^*} C - \frac{q_0 \gamma}{D^*} x\, e^{-\gamma t} + \frac{C_0 u}{D^*} = 0. \qquad (6.87)$$

Equation (6.87) is a linear differential equation of the first order. It is solved by ordinary methods [32]:

$$C(x, t) = - \frac{q_0 \gamma D^*}{u^2} \left(1 + \frac{xu}{D}\right) e^{-\gamma t} + \psi(t)\, e^{\frac{ux}{D^*}} + C_0, \qquad (6.88)$$

where $\psi(t)$ is some function of t.

Since at $x = 0$, $C = C_0$, then

$$\psi(t) = \frac{\gamma D^* q_0}{u^2} e^{-\gamma t}. \qquad (6.89)$$

Finally, for $C(x, t)$ we obtain the following solution, taking into account effective longitudinal diffusion:

$$C(x, t) = C_0 - \frac{q_0 \gamma}{u} x e^{-\gamma t} - \frac{q_0 \gamma D^*}{u^2} \left[e^{-\gamma t} - e^{-\gamma t + \frac{ux}{D}}\right]. \qquad (6.90)$$

From Eq. (6.90) it follows that a diffuse front of adsorbed substance is formed.

We may readily show that solution (6.90) changes to (6.84) if

$$\frac{D^* \gamma}{u^2} \ll 1. \qquad (6.91)$$

Longitudinal diffusion must thus be considered if the conditions of Eq. (6.91) are to be satisfied.

The evolved theory describes geochemical migration due to filtration in a weakly adsorbent medium if we know the adsorption capacity of the rock q_0, the coefficient of longitudinal diffusion D^* of the dissolved substance, and the rate constant of intragranular diffusion γ. These values may be determined experimentally. By using the solutions thus obtained, it is easy to compute the distance a substance will migrate, the amount that has migrated after a certain time, and the distribution law of the substance in the space above the deposit.

§ 47. Filtration in a Heterogeneous Medium with Consideration of Chemical Reaction

Let us examine the problem of the movement of a one-component solution (liquid or gaseous) through a porous medium with chemical reactions taking place between the substance and the medium. As pointed out in Chap. 4, the process in this case may take place in the kinetic, diffusion, and mixed regions.

Let a heterogeneous reaction between solution and rock or soil take place in the kinetic region. The migration process is described by the system of equations of material balance (1.17) and the kinetic equations of the appropriate chemical reaction for certain initial and boundary conditions. For simplicity we shall consider one-dimensional filtration of the solution along the x axis. The problem may be simplified if we consider that, because of the low diffusion rate, the amount of substance transferred along x by diffusion flow is normally much less than that transferred by filtration flow $(\vec{j}_D \ll \vec{j}_K)$, so that in the equation of material balance (6.76) written for the one-dimensional case, the member $D^* \partial^2 C / \partial x^2$ as compared with $u \partial C / \partial x$ may be neglected. In this case, the problem of one-dimensional filtration is described by a system of the equation of material balance

$$\frac{\partial C}{\partial t} + u \frac{\partial C}{\partial x} + \frac{\partial q}{\partial t} = 0 \qquad (6.92)$$

and the kinetic equation of the corresponding chemical reaction for definite initial and boundary conditions. Let us examine the solution of this problem for chemical reactions of different orders between substance and medium. We shall assume that the rate of heterogeneous chemical reaction in the kinetic range may be described by the equations of formal kinetics (Chaps. 4 and 5). This is valid if the reaction products do not retard the process.

1. **Irreversible Reaction of the Second Order.** The equation of rate is written in the form (5.45). The solution of the system of equations (6.92) and (5.45) for boundary conditions corresponding to continuous feeding of substance of constant concentration C_0 into the medium [condition (6.57)] was obtained in the paper of Panchenkov [33]; it has the form

$$C(x,\ t) = \begin{cases} \dfrac{C_0}{(e^{K' q_0 x / u} - 1) e^{K' C_0 \left(\frac{x}{u} - t\right)} + 1}, & x \leqslant ut, \\ 0, & x > ut, \end{cases} \qquad (6.93)$$

$$q(x,\ t) = \begin{cases} q_0 \dfrac{1 - e^{K' C_0 \left(\frac{x}{u} - t\right)}}{e^{K' C_0 \left(\frac{x}{u} - t\right) + K' q_0 \frac{x}{u}} - e^{K' C_0 \left(\frac{x}{u} - t\right)} + 1}, & x < ut, \\ 0, & x \geqslant ut. \end{cases} \qquad (6.94)$$

Equations (6.93) and (6.94) give a continuously diffuse front of reacting substance during its movement through the country rocks.

2. **Reversible Reaction of the First Order.** Equation (5.47) of the rate of first-order reversible reaction is analogous to Eq. (5.27) of the diffusion kinetics of adsorption for a linear isotherm $f(q) = q/K$ differing from it only by the value of the constant coefficient in front of the concentration gradient $C - q/K$. Therefore, to describe the migration of a substance in rocks or soils, we may use the solution (6.58) and (6.68) [in case of a continuous supply of solution with constant concentration to the medium, condition (6.57)] or (6.72) and (6.68) [for the initial and boundary conditions (6.71)] if in these solutions we substitute the value

$\gamma = \gamma_1\gamma_2/(\gamma_1 - \gamma_2)$ [see (5.28)] for the rate constant of the reaction K. K will here have the meaning of an equilibrium constant of the reaction (5.46).

3. **Reversible Reaction of the Second Order.** The kinetic equation has the form (5.50) if the reverse reaction takes place according to the monomolecular law, but the form (5.51) if the rate of the reverse process obeys the bimolecular law. The solution of the system of equations (6.92) and (5.50) for the initial and boundary conditions (6.57) was obtained by Thomas [35]; it may be written in the form

$$\frac{C}{C_0} = \frac{I_0\left(2\sqrt{ABxV}\right) + \varphi\left(\alpha V,\ \beta x\right)}{I_0\left(2\sqrt{ABxV}\right) + \varphi\left(\alpha V,\ \beta x\right) + \varphi\left(Bx,\ AV\right)},$$

$$\frac{q}{q_0} = \frac{\varphi\left(\alpha V,\ \beta x\right)}{I_0\left(2\sqrt{ABxV}\right) + \varphi\left(\alpha V,\ \beta x\right) + \varphi\left(Bx,\ AV\right)}, \tag{6.95}$$

where

$$A = \frac{K_2}{u}; \quad B = \frac{K_1 q_0}{u}; \quad \alpha = \frac{K_1 C_0 + K_2}{u};$$

$$\beta = \frac{1}{u}\frac{K_1 K_2 q_0}{K_2 + K_1 C_0}; \quad \varphi\left(u,\ V\right) = e^u \int_0^u e^{-t} I_0\left(2\sqrt{Vt}\right) dt, \tag{6.96}$$

where V is the volume of solution passing through the cross section x of the medium in time t, and I_0 is a Bessel function of the first kind of an imaginary independent variable [32].

A solution of the system of equations (6.92) and (5.51) for conditions (6.57) was also obtained by Thomas [35] and is analogous to the solutions (6.95) and (6.96).

Walter [36] obtained several partial solutions of the system (6.92) and (5.51) for conditions (6.57), and also asymptotic solutions (as time $t \to \infty$).

When $K_1 = K_2 = K$:

$$\frac{q}{q_0} = 1 - e^{-y} \int_0^\infty e^{-\tau} I_0\left(2\sqrt{x\tau}\right) d\tau, \tag{6.97}$$

where

$$y = \frac{q_0 K}{u} x; \quad t' = \frac{C_0 K}{u} V.$$

When $K_1 = K_2$:

$$\frac{q}{q_0} = \frac{1}{2}\left[1 + \operatorname{erf}\left(t'^{1/2} - y^{1/2}\right)\right]. \tag{6.98}$$

When $K_2 > K_1$:

$$\frac{q}{q_0} = \frac{1}{2}\left[1 + \operatorname{erf}\left\{\left(\frac{y}{K\left(1 + \beta\frac{q}{q_0}\right)}\right)^{1/2} - \sqrt{(1 - \beta q)\,y}\right\}\right], \tag{6.99}$$

where

$$K = \frac{K_1}{K_2}; \quad \beta = \frac{1 - K}{K}.$$

Reactions of higher order, as pointed out above (Chap. 4), are improbable, and they will not therefore be considered.

4. Chemisorption. The chemical adsorption of a substance by rocks may be considered an irreversible reaction (5.52). The problem is of interest to us because general solutions are obtained that may be used for describing types of migration of substances through rocks investigated below. The equation of chemisorption rate is written in the form (5.53). A solution of the system of equations (6.92) and (5.53) for conditions (6.57), describing geochemical migration in the case of a continuous supply of substance with constant concentration C_0 to the medium, was found by one of the authors of [37].

5. Irreversible Reaction Taking Place in the Diffusion Region. Let the reaction between substance and surrounding rock be irreversible, the reaction (5.44). If, as before, we assume that the transfer of substance by flow is greater than that by diffusion, geochemical migration is described by the system of equations (6.92) and (5.61) for definite initial and boundary conditions. It is easy to see that the kinetic equation (5.61) represents a particular case of Eq. (5.53) when $\varphi(q) = \gamma = $ const. Consequently, to describe the process of migration for initial and boundary conditions (6.57) we may use the solution of the problem of chemosorption dynamics [37], if we set $\varphi(q) = \gamma = $ const. It may be pointed out here that the expression for q (x, t) will have the form

$$
q(x, t) = \begin{cases}
0, & t \leq \dfrac{x}{u} \\[2ex]
\gamma C_0\left(t - \dfrac{x}{u}\right)e^{\dfrac{-\gamma}{u}x}, & \dfrac{x}{u} \leq t \leq \dfrac{x}{u} + \dfrac{q_0}{\gamma C_0}; \\[3ex]
q_0 l^{-\dfrac{\gamma}{u}\left[\left(1 + \dfrac{C_0}{q_0}\right)x - \dfrac{uC_0}{q_0}\left(t - \dfrac{q_0}{\gamma C_0}\right)\right]}, & \dfrac{x}{u} + \dfrac{q_0}{\gamma C_0} \leq t \leq \dfrac{x}{u}\left(1 + \dfrac{q_0}{C_0}\right) + \dfrac{q_0}{\gamma C_0}; \\[3ex]
q_0, & t \geq \dfrac{x}{u}\left(1 + \dfrac{q_0}{C_0}\right) + \dfrac{q_0}{\gamma C_0}
\end{cases}
\qquad (6.100)
$$

Let the reaction between mobile and immobile phases be reversible, reaction (5.46). It may be shown that the equation of diffusion kinetics for a reversible reaction of the first order agrees with Eq. (5.27) of adsorption kinetics in case of a linear isotherm $f(q) = q/K$, if by K we mean the equilibrium constant K_c of reaction (5.46). Therefore, to describe the migration of a substance from a steady source in the presence of first-order chemical reaction between substance and medium, we may use the solution (6.58) of adsorption dynamics.

It may be shown that the equation of diffusion kinetics of a second-order reversible reaction is analogous to the equation of adsorption kinetics for a convex isotherm. Consequently, the migration of a substance in this case conforms to the same laws observed for adsorption with a convex isotherm: a stationary front of dissolved substance is formed, moving through the medium at a rate indicated in Eq. (6.73).

§ 48. Filtration of a Mixture of Substances

The problem of filtration of a mixture of substances, taking into account interaction with the medium, is very complex, and no solution for the general case has yet been found. The difficulty of this problem may be illustrated by the example of filtration when adsorption takes place. If the adsorption isotherm is not linear, the value of adsorption of a given component

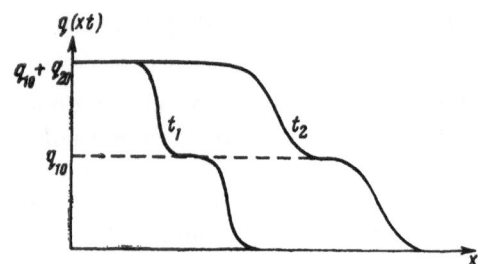

Fig. 30. Concentration distribution during continuous filtration of a mixture of two substances.

depends on the concentrations of the other components of the mixture [interdependent adsorption, Eq. (3.8)]. The system of equations of material balance and adsorption kinetics describing the filtration of a mixture in this case is so complex that it has been impossible to obtain an analytical solution.

Let the adsorption isotherm of each component of the mixture be linear:

$$q_i = K_i C_i, \qquad (6.101)$$

where q_i is the concentration of the i-th component (i = 1, 2, 3, ..., n, where n is the number of components of the mixture) in the mobile and immobile phases, respectively, and K is the adsorption coefficient of the i-th component.

Equation (6.101) is fulfilled for low concentrations of dissolved substance when each substance is adsorbed as if the other components were absent (interdependent adsorption). In this case the filtration of each substance in the medium will take place independently of the other. Therefore, the distribution of the i-th substance in the medium has the form (6.58) and (6.72) with consideration of Eqs. (6.61) and (6.68) depending on the initial and boundary conditions if we replace K by the adsorption coefficients K_i of the individual substances. We thus have a set of solutions, differing from each other by the values of K_i (i = 1, 2, 3, ..., n). The schematic picture observed during one-dimensional filtration of a mixture of dissolved substances is shown in Fig. 30 for continuous supply of a two-component solution with concentrations C_{10} and C_{20} to the medium; in Fig. 31 the picture is shown for the initial and boundary conditions (6.71). From Fig. 31 it follows that, beginning at some time, differential flow of the components of the system is observed. As a first approximation, it may be stated that each substance will move at its own characteristic rate:

$$V_i = \frac{u}{1 + K_i} \quad (i = 1, 2, ..., n). \qquad (6.102)$$

This conclusion agrees with the empirical principle of differential movement of elements in the earth's curst, formulated by Korzhinskii [38]. In particular, the removal of dissolved substance from the solvent during filtration — the so-called "filtration effect" in the terminology of Korzhinskii [39] — may also be due to adsorption or ion exchange in the interaction between solution and rock. The question of the relations of the so-called "filtration effect" and adsorption will be discussed in more detail below.

With interdependent adsorption of the mixture of substances, differences in adsorbability

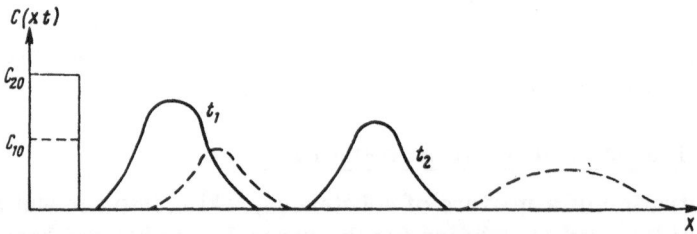

Fig. 31. Nonsharpening of band of two adsorbate substances during filtration.

determine the phenomenon of adsorption displacement: the substance with the greater adsorbability will displace and force aside the substance with lower adsorbability.

§ 49. A Dynamic Method of Determining the Kinetic Coefficients of Adsorption and Ion Exchange

From Eqs. (6.58), (6.64), and (6.70) it follows that, to describe the filtration of a substance through rocks, it is necessary to know two parameters: the effective coefficient of longitudinal diffusion D_{lo} and the adsorption coefficient K. These parameters may be found by means of laborious experiments on the study of kinetics and equilibrium of adsorption and ion exchange (Chaps. 3 and 4). We shall show that D_{lo} and K are readily determined by a simple dynamic experiment from the effluent curve of frontal analysis (i.e., from the time dependence curve of solution concentration at the outlet of a column supplied by a continuous feed of substance).

This is of great practical significance, since the adsorption (ion-exchange) rate is determined from a dynamic experiment that is less laborious than the direct experimental study of kinetics. Thus, Ekedahl, Högfeldt, and Sillén [40] have stated that, whereas the kinetic experiment on exchange takes something on the order of three months, the same results may be obtained from a dynamic experiment in 10 hours. On the basis of Eq. (6.62) it is easy to show that

$$K = \frac{u T_{1/2}}{l} - 1,$$
(6.103)

where $T_{1/2}$ is the time for the point of half concentration to appear, and $C = C_0/2$ at the outlet of a column of length l.

By determining $T_{1/2}$ experimentally, it is possible to compute the adsorption coefficient K in accordance with Eq. (6.103).

For finding D_{lo} let us write the equation of the effluent curve of frontal analysis, which follows from expression (6.58):

$$C(l, t) = \frac{C_0}{2}\left[1 - \text{erf}\left(\frac{l - \frac{u}{1+K}t}{2\sqrt{\frac{D_{lo}}{1+K}t}}\right)\right].$$
(6.104)

The values of the Gaussian integral [32] are obtained from graphs and tables. Having taken the definite ratio $C(l, t)/C_0$ from the experimental effluent curve, we may, by using graphs and tables, find the independent variable z of the Gaussian integral, which in this case is equal to

$$z = \frac{l - \frac{u}{1+K}t}{2\sqrt{\frac{D_{lo}}{1+K}t}}.$$
(6.105)

The values of l and t are known from experiment; K is determined from Eq. (6.103), and the effective coefficient of longitudinal diffusion is found from the expression

$$D_{lo} = \left(\frac{l - \frac{u}{1+K}t}{2z}\right)\frac{1+K}{t}.$$
(6.106)

§ 50. Experimental Results on Filtration of Solutions Not Interacting with the Rocks

The questions of filtration of salt solutions through porous media have been widely investigated experimentally in connection with reclamation of water-logged land, hydraulic engineering, and the like. Theoretical discussions of the results of investigating filtration of salts through columns of quartz sand have appeared in a number of papers [13, 41-43].

We shall consider the results obtained in those investigations in which radioactive tracers were used to study the filtration of KH_2PO_4 (with the radioactive tracer P^{32}) through a column filled with quartz sand (the fraction 0.25-0.5 mm). The sorption of cations on quartz sand, although it does take place [44], is so slight that the interaction between KH_2PO_4 and the medium may be neglected. The experiments were conducted for filtration of a solution through a column initially free from salt solution [conditions (6.47)] and for the elution of salt solution from the column. Asymptotic solutions of the problems of filtration and elution of solutions from columns have the form of Eqs. (6.36) and (6.46). Their properties have been analyzed in Eqs. (6.41)-(6.44).

Experiments have established the fact that a diffuse dynamic front is formed, the length of which increases with time $\sim\sqrt{t}$, which is in agreement with theory [Eq. (6.44)]. Of all concentration points, only one, the point of half concentration ($C = 0.5C_0$), moves at the constant rate u, which agrees with Eq. (6.42). By measuring the length of the front experimentally, the effective coefficient of longitudinal diffusion D_{lo} was computed from Eq. (6.44). Depending on the filtration rate and the concentration, D_{lo} ranged between wide limits, from 0.01 to 1 cm^2/min. This value is more than an order higher than the diffusion coefficient of KH_2PO_4 in water. Consequently, the spreading of the front of dissolved substance during filtration is determined chiefly by the inhomogeneity of the medium.

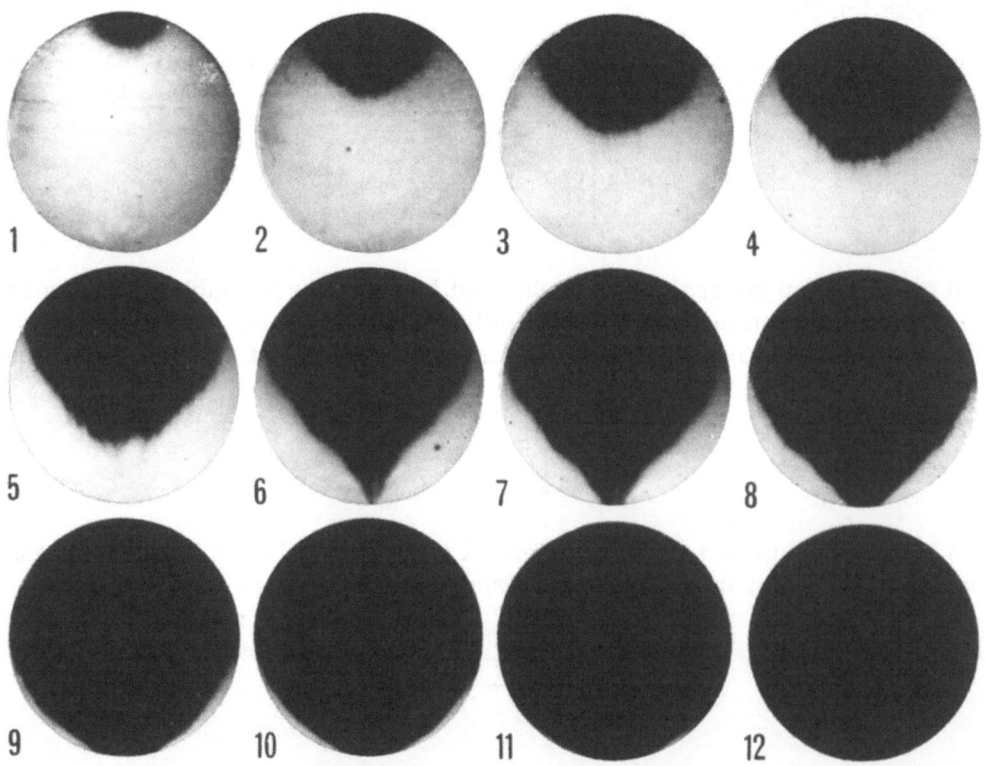

Fig. 32. Sequence of filling of the filter by solution.

On the basis of the investigations, it has been concluded [13, 41–43] that the theory of filtration of solutions through the column is in agreement with experiment. Since theoretical equations of filtration of solutions through a column [13] asymptotically coincide with Eqs. (6.36) and (6.46), which describe the filtration of solutions through rocks from a semiinfinite bed or from a steady source, it may be assumed that Eqs. (6.36) and (6.46) also agree with experiment. Consequently, by means of Eqs. (6.39) and (6.40) we may compute the distance of geochemical migration due to filtration.

We should note, however, that the theory considered in preceding paragraphs of this section and supported by several investigators [13, 41–43] does not account for interaction between solutions and rocks, and, as a rule, this is not in agreement with natural conditions of migration. A theory, considering adsorption, ion exchange, and chemical reactions between dissolved substances and rocks, has been developed above. Its application in practical computations of geochemical migration will be shown in Chap. 7.

It was pointed out above that when filtration is nonlinear, the distribution of a solution within the filter and the effluent curves depend on the shape of the filter and the methods of supplying the solution and removing the filtrate from the filter. The authors conducted a number of experiments [23] for the purpose of illustrating the indicated dependence. These questions have been inadequately covered in the literature, but they are important in geology since they have a direct relation to the complex paths along which natural solutions pass.

A cylindrical filter 24 cm in diameter and 0.5 cm high was prepared. The frame of the filter was made of Plexiglas and had two small openings at diametrically opposite points on the sides. The filter was filled with quartz sand having grain diameters of about 1 mm. The experiment was carried out according to the scheme of frontal analysis. A stream of water was followed immediately by a stream of 1% potassium permanganate solution at a rate of 40 ml/min. Through the transparent Plexiglas we visually observed the character of solution flow into the filter. It was established experimentally that the pore volume of the filter, approximately equal to 90 ml, was completely filled with the potassium permanganate solution only after about 150 ml of solution had been fed to the filter. The first twelve stages of filling the filter with solution, on the basis that each new stage corresponds to successive additions of 10 ml of solution fed to the filter, are recorded photographically (Fig. 32). From this figure we may see that at the times of stages 6–11 the potassium permanganate solution was diluted with water. The phenomenon of dilution, due to irregular paths along which the solution passed, determines to a considerable degree the shape of the effluent curve.

§ 51. A Study of Diffusion and Filtration of Adsorbable Solutions and Gases in Rocks

It should be noted that very little experimental work on studying diffusion and filtration of solutions and gases in rocks has been interpreted by means of the theory considered in the present chapter, nor in many works not connected with geochemical subjects. This is not remarkable in view of the fact that well developed theories are new in geochemistry.

Evans and Barber [45] investigated the diffusion of RbCl in different soils and clays by means of radioactive tracers (Rb^{86}). Those investigators learned that Rb^+ ions from the solution are exchanged with ions in the rocks. For treating the results of measurements they therefore used not the solution of Eq. (2.4) of Fick diffusion [with initial and boundary conditions appropriate to the setup of the experiment, in the present case with conditions (2.32)] but the solution (6.16) of the diffusion problem considering ion exchange. The computed effective diffusion coefficient of RbCl in kaolin changes according to the content and is on the order of D_{ef} $(1.2-5.7) \cdot 10^{-9}$ cm^2/sec. This value is much less than the diffusion coefficient of RbCl in solution. Consequently, ion exchange leads to a decrease in the diffusion rate of the exchanging ion. It must be noted that such small diffusion coefficients as RbCl has in kaolin are probably due not only to ion exchange but also to incomplete wetting of the soil.

Briling[46] studied the diffusion of calcium chloride in morainic carbonate-free clay of the Moscow region by the method of stationary flow. The exchange capacity of the clay was found to be 31 mg-equiv per 100 g of dry material. Of this capacity, 1.0 mg-equiv belongs to sodium, 18 to magnesium, and 12 to calcium. Figure 33 shows a typical picture of time dependence of ions diffused through a layer of clay. It may be seen that the amount of Ca^{2+} diffused in a definite time is less than the amount of Cl^-. This is explained by ion exchange, as a result of which some of the Ca^{2+} is adsorbed by the clay and an equivalent amount of Mg^{2+} and Na^+ is removed from the clay to the solution, diffusing in it. To treat the results of his measurements, Briling used an equation of adsorption kinetics analogous to Eqs. (6.24) and (6.31). However, integration of the kinetic equation was not altogether accurately performed, and the physical meaning of the kinetic factor in front of the concentration gradient, $C - f(q)$, has not been interpreted. The following results were obtained in the investigated work: at moisture contents of 15, 17, and 20%, the diffusion coefficients were found to be, respectively, 0.09, 0.106–0.109, and 0.113 cm^2/day. Briling [46] also derived an equation describing stationary diffusion with consideration of adsorption by the rocks (a problem we have not investigated).

From the experimental data of Briling [46], it follows that the kinetics of diffusion may be described by an equation of the type of (6.24) and (6.31).

Antonov [47, 48], whose work was examined in detail in § 12 of Chap. 3, studied the diffusion of hydrocarbon gases in rocks. However, he did not substantiate the possibility of using Fick's laws of diffusion to describe the diffusion of adsorbable gases. Let us examine the problem of diffusion of a gas in rocks. We shall designate the concentration of gas dissolved in the pore water by C_w, the concentration of gas in the free space of the pores by C_{fr}, and the concentration of adsorbed gas by q. For one-dimensional diffusion of gas in the rock, the system of equations (6.1) and (6.3) is then written in the form

$$\frac{\partial C_w}{\partial t} + \frac{\partial C_{fr}}{\partial t} + \frac{\partial q}{\partial t} = D_w \frac{\partial^2 C_w}{\partial x^2} + D_{fr}\frac{\partial^2 C_{fr}}{\partial x^2}, \qquad (1.107)$$

$$\frac{\partial q}{\partial t} = K_1 \frac{\partial C_{fr}}{\partial t}, \qquad (1.108)$$

where D_w and D_{fr} are the diffusion coefficients for the dissolved gas and in the pore space, and K_1 is the adsorption coefficient of gas in the pore space. Since $C_w = \gamma C_{fr}$ [γ is the solubility coefficient of gas in water, in accordance with Eq. (2.112)], by substituting the expression in (6.108) in Eq. (6.107), we obtain

$$\frac{\partial C_{fr}}{\partial t} = D_{ef}\frac{\partial^2 C_{fr}}{\partial x^2}, \qquad (6.109)$$

$$D_{ef} = \frac{D_{fr} + D_{w}\, \gamma}{1 + \gamma + K_1}. \qquad (6.110)$$

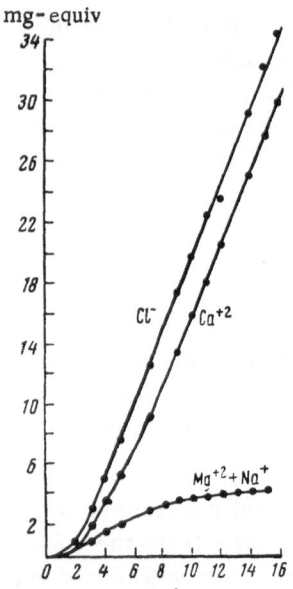

Fig. 33. Time dependence of the number of ions diffused through a layer of clay.

On comparing Eqs. (6.110) and (2.4) it is seen that we may use Fick's diffusion laws for describing the diffusion of gas through rocks, but the diffusion coefficient loses its ordinary meaning and becomes an effective value, determined from Eq. (6.110). In deriving this equation, the adsorption of dissolved gas was not accounted for. If we do account for it, we obtain

$$D_{ef} = \frac{D_{fr} + D_w\,\gamma}{1 + \gamma + K_1 + K_2\gamma}, \tag{6.111}$$

where K_2 is the adsorption coefficient of gas dissolved in water.

Experiments on the filtration of dissolved substances and gas in rocks have not been numerous. In a number of papers [49-52], the dynamics of ion-exchange of salts in soils has been discussed on the basis of studies made to explain the nature of solution movement through soils.

Solutions containing one or several ions (Cu^{2+}, Co^{2+}, Rb^{2+}, Hg^{2+}) have been passed through columns filled with soil. It has been established qualitatively that the ions become fractionated during their passage through the column [49].

Rachinskii and Fokin [51] studied the sorption of phosphates in soil (heavy loamy podsol) by means of radioactive tracers. The adsorption isotherm of the phosphates was convex. In agreement with theory [Eq. (6.73)] Rachinskii and Fokin [51] observed a stationary front of adsorbed substance during filtration through the soil column. Treatment of the experimental data and comparison of them with results obtained earlier [13] showed that the diffuseness of the phosphate front in the soil column is due to longitudinal diffusion and to hydrodynamic factors. This is due to the fact that the flow rate was low, 10^{-4} cm/sec, as a consequence of which equilibrium between adsorbate and adsorbent could be established for each section of the column at any time.

Similar results may be arrived at from analysis of kinetic data that we obtained on the exchange of Cu^{2+} and H^+ ions on bentonite. As we showed in Chap. 5, the rates of intergranular and intragranular diffusion (at flow rates used in the experiments, $u \approx 0.015\text{-}0.15$ cm/min) are greater than the feed rate of substance to the adsorbent by flow. In the bentonite layer one grain thick, adsorption equilibrium obtains at any instant. Consequently, when Cu^{2+} and H^+ ions are filtering through a column with bentonite, the front will be spread only because of longitudinal diffusion and hydrodynamic factors.

§ 52. The So-called "Filtration Effect"

According to the terminology of Korzhinskii [39], "filtration effect" means the lagging behind of dissolved substance in a solvent during movement through porous media. The theory of the "filtration effect" was formulated by Korzhinskii by introducing a filtration coefficient φ, which is equal to the ratio of the rate of movement of dissolved substance to the rate of the solvent and by taking into account the material balance of the dissolved substance. However, Korzhinskii derived only differential equations describing the so-called "filtration effect." No analytical solutions of these equations have been supplied. According to Korzhinskii and Zharikov [54, 55], the physicochemical nature of the "filtration effect" lies hidden behind the coefficient φ.

To prove the "filtration effect," the filtration of solutions of the electrolytes $CuCl_2$, $CuSO_4$, $FeCl_3$, $Fe_2(SO_4)_3$ and others through columns filled with quartz powder of certain fractions (0.02-0.025, 0.015-0.019, 0.009-0.014, and 0.002-0.008 mm) were studied [53, 54]. In these studies, the concentration of introduced cations at the outlet of the column proved to be lower in the first samples than in the initial solution. Zharikov, Dyuzhinova, and Maksakova [53, 54] explain this by the "filtration effect," in which the flow rates of cations and anions differ. The results obtained may be explained by ion exchange between cations in the solution and the hydrogen ions on the surface of quartz interact with water, giving rise to silicic acid. The latter dissociates in accordance with the equation

$$H_2SiO_3 \rightleftharpoons 2H^+ + SiO_3^{-2}.$$

Silicate ions, SiO_3^{2-}, being less mobile, remain on the surface, and the particle thus acquires a negative charge. Hydrogen ions are arranged near the surface. Since the cations fed into the column are adsorbed, then, in accordance with the theory of adsorption dynamics (Chap. 6), the cation front lags behind the solvent. The dependence of the filtration coefficient φ of $CuCl_2$ and $FeCl_3$ on various parameters of the experiment, as found by Zharikov, Dyuzhinova, and Maksakova [53, 54], may be explained on the basis of the theory of adsorption dynamics and ion exchange.

Let us consider the dependence of the "filtration effect" on solution concentration. It is known [44] that the adsorption isotherm of copper on quartz powder is convex. As follows from the theory of adsorption dynamics, a stationary front is formed during the filtration of copper through quartz powder, moving through the column at a rate indicated in Eq. (6.73). Consequently, the semipermeability coefficiently, the semipermeability coefficient α is equal to

$$\alpha = 1 - \varphi = 1 - \frac{v}{u} = \frac{q_0}{q_0 + C_0}, \qquad (6.112)$$

where v and u are the rates of movement of dissolved substance and solvent, respectively. From this equation, it follows that the semipermeability coefficient α declines with increase in concentration C_0. But α does not depend on the mass of quartz powder (length of column). All this is in agreement with experimental results on $CuCl_2$ and $FeCl_3$ [53, 54].

To investigate the nature of the "filtration effect" in quartz sand, we conducted a number of experiments on the filtration of a solution of $CuCl_2$ through a column filled with sand [23]. In one of the experiments a solution of $CuCl_2$ of 0.03 N concentration was fed immediately after a flow of water into a rubber tube 10 m long and 12 mm in diameter, filled with sand. The tube filling was quartz sand from the Lyubertsy pit, from which the fraction 0.2-0.4 mm was sieved off and was washed in boiling hydrochloric acid before using. The effluent solution from the column was analyzed for its cooper and chlorine contents by iodometric and argentometric titration. The filtration rate was on the order of 20 ml per hour. The results of the experiment in the form of time-dependent curves of Cu^{2+} and Cl^- concentrations at the outlet of the column are shown in Fig. 34. Investigations of the adsorption of copper ions on quartz sand were carried out in the same column. Into the column, which was first filled with water, 50 ml of $CuCl_2$ with a concentration of 0.03 N was introduced, and the column was then flushed with

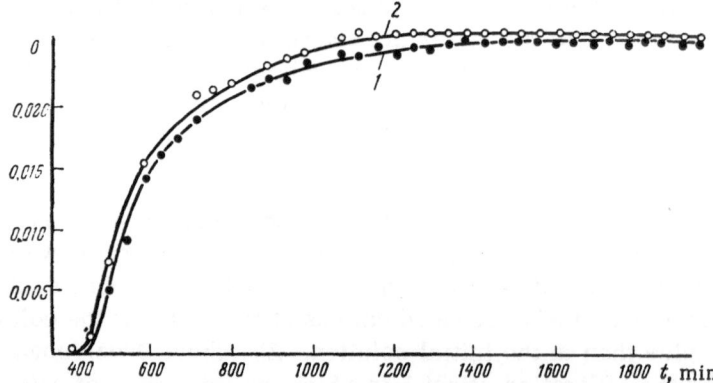

Fig. 34. Effluent curves of a column for copper (1) and
chlorine (2) ions for frontal analysis.

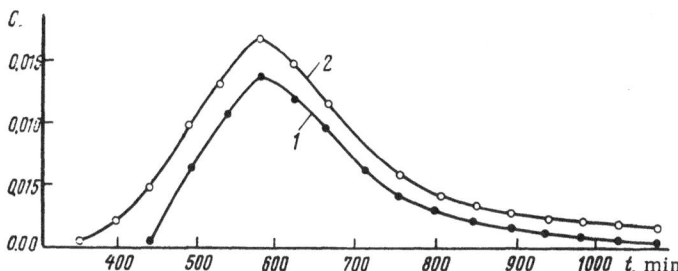

Fig. 35. Effluent curves of copper (1) and chlorine (2)
ions during elution of a band of $CuCl^2$ through a column.

distilled water. The effluent solution from the column was analyzed for the contents of copper
and chlorine ions. The time dependence of the Cu^{2+} and Cl^- concentrations at the outlet is
shown in Fig. 35.

In Fig. 34 it is seen that Cl^- ions actually outstrip the Cu^{2+} ions somewhat; i.e., a "filtration effect" is observed, similar to that in [53, 54]. From Fig. 35 it follows that copper ions
are adsorbed on quartz sand, the value of the adsorption capacity, according to our data, being
at least $2 \cdot 10^{-5}$ g of Cu^{2+} per gram of sand. There is no adsorption of the chlorine ion. Consequently, on the basis of experimental data, we may conclude that the lagging of copper ions
behind chlorine ions at the outlet of the column (Fig. 34) is due to the adsorption of copper ions
by quartz sand. The fractionation of cations and anions during filtration, due to the exchange
of ions between solution and rock, does not depend on the size of the pores but is determined by
the surface area of the rock. It was for this reason that it was found possible to produce the
so-called "filtration effect" on ordinary natural sand but not on fine quartz powder. The experimental results obtained confirm the view expressed above concerning the adsorptive nature
of the "filtration effect" observed by Zharikov, Dyuzhinova, and Maksakova [53, 54].

These three investigators, in proposing an electrokinetic character to the "filtration
effect" in quartz sand, measured the adsorption of electrolyte solutions in quartz powder by a
static method and concluded that iron is not adsorbed and that the adsorption of copper is small.
This conclusion [54] that the "observed change in concentration of solutions in the filtration experiments cannot be caused by adsorption of the dissolved components by quartz" is without
adequate foundation, since one cannot compare the results of static and dynamic experiments
on adsorption, even if they are made with identical relations between mass of solution and
quartz powder. Actually, if ions are exchanged between the surface of the quartz layer and the
solution, then, under static conditions, this exchange will proceed until equilibrium is established, but in a current it will go farther, to saturation of the surface layer with ions from the
solution, and, consequently, the amount of adsorbed substance in the dynamic experiment will be
greater than in the static. Zharikov and his coworkers [53, 54] believe that the dependence of
the "filtration effect" of $CuCl_2$ and $FeCl_3$ in quartz powder on various parameters of the experiment contradicts the adsorption laws. However, the reverse was shown above: the dependence
of the "semipermeability coefficient" α on the solution concentration and its independence of the
mass of quartz powder support the adsorptive nature of the "filtration effect" of electrolytes
in quartz. Electrokinetic phenomena during filtration were observed by Ovchinnikov and Shur
[56, 57] during ultrafiltration (filtration under the effect of a large pressure gradient) through
solid filters. A "filtration effect" due to electrokinetic phenomena was not observed by Zharikov and his coworkers [54, 55] because they did not set up their experiments for ultrafiltration, but rather for frontal chromatography on quartz powder.

In conclusion we shall discuss the relation between the theory of so-called "filtration effect" [58] for the particular case of one-dimensional filtration of a single-component solution (along the x axis) and the quantitative theory of heterogeneous processes of geochemical migration [the system (1.9) and (1.17)]. According to the theory of "filtration effect" [58], the one-dimensional flow of dissolved substance along the x axis is described by the equation

$$\frac{\partial q}{\partial t} + v\,\frac{\partial C}{\partial x} = D\,\frac{\partial^2 C}{\partial x^2}\,, \tag{6.113}$$

where v is the rate of movement of the dissolved substance,

$$v = \varphi u, \tag{6.114}$$

and φ is the filtration coefficient.

Let us compare Eq. (6.113) with the equation of material balance (1.17), which for the one-dimensional case has the form

$$\frac{\partial q}{\partial t} + u\,\frac{\partial C}{\partial x} + \frac{\partial C}{\partial t} = D\,\frac{\partial^2 C}{\partial x^2}\,. \tag{6.115}$$

Since

$$\frac{\partial C}{\partial t} = \frac{\partial C}{\partial x}\left(\frac{dx}{dt}\right)_C, $$

then, in place of Eq. (6.115), we obtain

$$\frac{\partial q}{\partial t} + \left[u + \left(\frac{dx}{dt}\right)_C\right]\frac{\partial C}{\partial X} = D\,\frac{\partial^2 C}{\partial X^2}\,. \tag{6.116}$$

On comparing Eqs. (6.116) and (6.117) it may be seen that

$$v = u + \left(\frac{dx}{dt}\right)_C. \tag{6.117}$$

The value $(dx/dt)_C$ is the rate of movement of the concentration point C = const of the front. If $(dx/dt)_C$ = const, i.e., if the front of dissolved substance moves at constant rate (stationary front), then v is not equal to v (C), and it has a definite physical significance [in accordance with Eq. (6.117)]. For spreading dynamic fronts $(dx/dt)_C \neq$ const and the value v has no clear physical meaning.

Thus, the concept of the so-called "filtration effect" may be used if a stationary dynamic front of dissolved substance is established during filtration. This is observed (as indicated above) during filtration in an adsorbent medium, when the adsorption isotherm is convex. The presence of a stationary front, however, may be established only by solving the system of equations (1.9) and (1.17) for certain initial and boundary conditions.

Equation (6.113) is considered by Zharikov [58] to be a generalized equation of infiltration and diffusion metasomatism. From the above discussion, it follows that on the basis of Eq. (6.113) [to which we must add Eq. (1.9), since the first contains two unknown functions C and q] we cannot obtain a quantitative solution of the problem of geochemical migration. Such solutions may be found only by using the system of equations (1.9) and (1.17).

LITERATURE CITED

1. Crank, J., The Mathematics of Diffusion, Oxford (1956).
2. Wagner, C., J. Chem. Phys. Vol. 18, p. 1229 (1950).
3. Stokes, R. H., Trans. Faraday Soc., Vol. 48, p. 887 (1952).
4. Fujita, H., Textile Res. Inst. Vol. 22, pp. 757, 823 (1952).
5. Timofeev, D. P., The Kinetics of Adsorption [in Russian] (1962).
6. Rozenshtok, Yu. L., Zh. Fiz. Khim., Vol. 39, p. 1135 (1965).
7. Dubov, R. I., in: Geochemistry of Ore Deposits [in Russian] (1964).
8. Danckwerts, P. V., Trans. Faraday Soc., Vol. 47, p. 1014 (1951).
9. Wilson, A. H., Phil. Mag., Vol. 39, p. 48 (1948).
10. Crank, J., Phil. Mag., Vol. 39, p. 362 (1948).
11. Katz, S. M., Kub, E. T., and Wakelin, I. N., Textile Research Inst., Vol. 20, p. 754 (1950).
12. Reese, C. E., and Eyring, H., Textile Research Inst., Vol. 20, p. 743 (1950).
13. Rachinskii, V. V., Chia Ta-ling, and Chistova, E. D., Izv. TSKhA, No. 2, p. 105 (1962).
14. Boltaks, B. I., Diffusion in Semiconductors, Academic Press, New York (1963).
15. Gertsriken, S. D., and Dekhtyar, I. Ya., Diffusion in Metals and Alloys in the Solid Phase [in Russian] (1966).
16. Patrashev, A. N., and Aratyunyan, N. K., Izv. NII Gidrotekhniki, Vol. 30, p. 64 (1941).
17. Patrashev, A. N., Izv. NII Gidrotekhniki, Vol. 31, p. 55 (1946).
18. Aldul, S. P., in: Questions on the Formation of the Chemical Composition of Ground Water [in Russian], p. 104 (1963).
19. Glueckauf, E., Trans. Faraday Soc., Vol. 51, p. 34 (1955).
20. Helfferich, F., Ion Exchange, McGraw-Hill, New York (1962).
21. Golubev, V. S., Dissertation, Moscow University (1964).
22. Zhukhovitskii, A. A., and Turkel'taub, N. M., Gas Chromatography [in Russian] (1962).
23. Garibyants, A. A., Golubev, V. S., and Beus, A. A., Izv. Akad. Nauk SSSR, Ser. Geolog., No. 9, p. 26 (1966).
24. Tikhonov, A. N., Zhukhovitskii, A. A., and Zabezhinskii, Ya. L., Zh. Fiz. Khim., Vol. 20, p. 1113 (1946).
25. Todes, O. M., Zh. Fiz. Khim., Vol. 18, p. 591 (1945).
26. Todes, O. M., and Bikson, Ya. M., Dokl. Akad. Nauk SSSR, Vol. 75, p. 727 (1950).
27. Bikson, Ya. M., Zh. Fiz. Khim., Vol. 27, p. 1530 (1953); Vol. 28, p. 1017 (1954).
28. Todes, O. M., and Rachinskii, V. V., Zh. Fiz. Khim., Vol. 29, pp. 1591, 1909 (1955).
29. Rachinskii, V. V., and Todes, O. M., Zh. Fiz. Khim., Vol. 30, p. 407 (1956).
30. Rachinskii, V. V., Zh. Fiz. Khim., Vol. 39, p. 444 (1957); Vol. 36, p. 2018 (1962).
31. Golubev, V. S. and Panchenkov, G. M., Zh. Fiz. Khim., Vol. 37, p. 1010 (1964).
32. Bronshtein, I. N., and Semendyaev, K. A., Handbook of Mathematics [in Russian], Izd. Fiz. -Mat. Lit., Moscow (1959).
33. Panchenkov, G. M., Zh. Fiz. Khim., Vol. 38, p. 770 (1964).
34. Thomas, H. S., in: Chromatography [Russian translation], Izd. Inostr. Lit. (1949).
35. Thomas, H. C., J. Am. Chem. Soc., Vol. 66, p. 1664 (1944).
36. Walter, J. E., J. Chem. Phys., Vol. 13 p. 332 (1945).
37. Golubev, V. S., Kuz'min, E. N., and Panchenkov, G. M., Zh. Fiz. Khim. Vol. 39, p. 1018 (1965).
38. Korzhinskii, D. S., Zap. Vserossiisk. Mineralogich. Obshch., Vol. 71, p. 160 (1942).
39. Korzhinskii, D. S., Izv. Akad. Nauk SSSR, Ser. Geolog., No. 2, p. 35 (1947).
40. Ekedahl, E., Högfeldt, E., and Sillén, L., Nature, Vol. 166, p. 723 (1950).
41. Rachinskii, V. V., Chia Ta-ling, and Chistova, E. D., Izv. TSKhA, No. 1 (1963).
42. Rachinskii, V. V., and Asanakunov, A., Dokl. TSKhA, No. 99, p. 533 (1964).
43. Rachinskii, V. V., and Asanakunov, A., Izv. TSKhA, No. 2, p. 186 (1965).
44. Heydemann, A., Geochim. et Cosmochim. Acta, No. 15, p. 305 (1959).

45. Evans, S. D., and Barber, S. A., Soil. Sci. Soc. Proc., No. 28, p. 53 (1964).

46. Briling, I. A., Vestn. MGU, No. 5, p. 62 (1965).

47. Antonov, P. L., Tr. NII Geofiz. i Geokhim. Metodov Razvedki, No. 2 (1957).

48. Antonov, P. L., "Direct methods of prospecting for oil and gas," Tr. NII Geofiz. i Geokhim. Metodov Razvedki (1964).

49. Gapon, E., and Chernikova, T., Dokl. TSKhA, No. 7, p. 26 ((1948).

50. Antipov-Karataev, M. N., Paskvik-Khlopina, M. A., Merkulova, M. S., and Grebenshchikova, V. I., Kolloid. Zh., No. 10, p. 400 (1948).

51. Rachinskii, V. V., and Fokin, A. D., Dokl. TSKhA, No. 94, p. 33 (1963).

52. Thomas, H. C., Agricultural and Food Chemistry, No. 11, p. 201 (1963).

53. Zharikov, V. A., Dyuzhinova, T. N., and Maksakova, É. M., Izv. Akad. Nauk SSSR, Ser. Geolog., No. 1, p. 41 (1962).

54. Zharikov, V. A., Dyuzhinova, T. N., and Maksakova, É. M., Izv. Akad. Nauk SSSR, Ser. Geolog., No. 10, p. 81 (1963).

55. Zharikov, V. A., in: Problems of Geochemistry [in Russian], Izd. AN SSSR, p. 276 (1965).

56. Ovchinnikov, L. N., and Shur, A. S., in: Transactions of the Conference on Experimental Mineralogy and Petrography (Tr. Soveshchaniya po Eksperiment. Mineralogii i Petrografii) [in Russian], Izd. AN SSSR, p. 163 (1953).

57. Shur, A. S., and Popov, G. P., in Physicochemical Problems in the Formation of Rocks and Ores (Fiziko-khim. Problemy Formirovaniya Gornykh Porod i Rud) [in Russian], Izd. AN SSSR, p. 654 (1961).

58. Zharikov, V. A., Geokhimiya, No. 10, p. 1191 (1965).

THE THEORY OF FORMATION OF HYDROTHERMAL DEPOSITS AND GEOCHEMICAL AUREOLES AT DEPOSITS OF ORES AND GAS

By geochemical aureoles we mean zones of high concentrations of elements in rocks, plants, ground water and gas near mineral deposits, with which these zones are genetically related. The formation of geochemical aureoles of different types takes place chiefly by filtration of ore-forming solutions and gases, the diffusion of dissolved substances, and by physical weathering and mechanical transfer of substances of the deposit [1, 2], which is accompanied by interaction between migrating substances and rocks. The principal processes of interaction of solutions and gases with the country rock are adsorption, ion exchange, and chemical reactions.

Hydrothermal deposits and their primary aureoles are formed in the following basic ways: 1) by reaction between hydrothermal solutions and the country rocks, 2) by crystallization as a result of changes in the thermodynamic conditions of migration of an ore-forming solution, 3) by chemical reactions between components of the hydrothermal solutions with the formation of insoluble compounds.

There are different ways of classifying geochemical aureoles according to the principal feature on which we base the classification. Depending on time of formation, we distinguish primary aureoles, which formed at the same time as the deposit and as a result of the same genetic processes, and secondary aureoles, due to dissemination, both of the deposit and of its primary aureole. Depending on the phase in which the migrating substance is found, we may distinguish lithochemical, hydrochemical, and atmochemical aureoles. Among lithochemical aureoles, we may distinguish mechanical and salt varieties. Mechanical dissemination aureoles are formed by mechanical destruction of rocks and included ore bodies that are exposed at the surface [1]. Ore particles that form in this process migrate under the influence of external forces. Salt aureoles are generally formed by diffusion of salt in a water-soaked medium, accompanied by sorption and chemical reactions of varying magnitude with substances in the rocks (diffusion aureole). In the formation of gas aureoles, the principal role is played by gas diffusion [3], adsorption, and chemical reactions between the diffusing gases and the rocks. The filtration of gases and liquids in bedrock under the pressure gradients existing in nature is possible only along faults or other fractures [3]. Primary geochemical aureoles are formed by filtration along fractures. The role of filtration in the formation of secondary diffusion aureoles reduces chiefly to a distortion of the continuous distributions of substances above the deposit because of diffusion. There are also biochemical aureoles, in the formation of which plants and microorganisms participate. We shall not consider this type of geochemical aureole.

Each classification of geochemical aureoles has its advantages and disadvantages. From the theory of geochemical migration developed in the previous pages, it follows in particular, that equations describing migration of a substance in the liquid and gaseous phases match each other in a number of cases. It is therefore more suitable to divide geochemical aureoles into primary and secondary.

In 1964 Dubov [4, 5] attempted a quantitative description of the formation of geochemical aureoles on the basis of diffusion and filtration laws. In his paper [5] he solved the problem of one-dimensional diffusion without consideration of adsorption, which is inadequate for explaining the processes of primary aureole formation. In a section on "Diffusion in the presence of sorption," the author examined the problem of one-dimensional diffusion with consideration of irreversible chemical reactions of the first order between the diffusing substance and the medium (see Chap. 6). The indicated solution does not take into account the mechanism of reversible adsorption, since the author did not use an equation of the adsorption isotherm in solving the problem but rather an equation of the kinetics of a first-order irreversible reaction. Neither did he consider the most general case of formation of primary geochemical aureoles during filtration of mineral-forming solutions in which chemical reactions of a different order were taken into account.

Below, on the basis of the theory of heterogeneous processes of geochemical migration developed in the preceding chapters, we shall examine some dynamic models of the formation of geochemical aureoles and hydrothermal deposits.

§ 53. The Formation of Hydrothermal Deposits and Primary Aureoles by Interaction between Solution and Country Rock

In the simplest case, a primary geochemical aureole of an ore deposit or gas deposit may be represented as a zone of alteration, resulting from interaction between ore solutions or gases and rocks surrounding fractures along which the filtration of the ore-bearing solutions and gases has taken place. Therefore, below we shall examine the problem of filtration of a solution interacting with the walls of an infinitely extended fracture.

Formulation of the Problem. Let us consider the simplest case. Let there be an infinitely extended macrofracture with plane-parallel walls separated by the distance l, and let a one-component solution (liquid or gaseous) filter through this fracture at a constant rate u, interacting with the walls (adsorption by the walls, exchanging ions, or entering into chemical reactions with the minerals of the country rock). It is assumed that the country rock contains microfractures filled with an aqueous solution that is in equilibrium with the country rock. For simplicity we shall assume that the microfractures are isotropically arranged in space. The mineral-forming hydrothermal solution or gas diffuses into the microfractures, emerging on the surface of the walls. Filtration in the microfractures may be neglected, since a large pressure gradient is necessary for this, the existence of which is unlikely in nature.

Let us place the origin of the coordinate system in the middle of the fracture, the x axis directed along the fracture, y perpendicular to the walls. We shall designate the initial concentration of ore solution by C_0, and indicate the rate of movement of this solution along the fracture by u. Let us assume that in the zone x < 0 the hydrothermal solution weakly interacts with the surrounding rock (which may be due to the chemical composition of the rock or to low porosity); this interaction will not be taken into account. Above the section x = 0 the ore-bearing solution begins to interact with the rocks, forming an ore body. Thus, the lower boundary of the ore body in the adopted model is tentatively assumed to be sharp. We are required to determine the distribution law of the concentration for the dissolved substance and the substance formed by reaction in the macrofracture and in the surrounding rocks at any instant in time.

Solution of the Problem. To solve the problem we have formulated we shall set up an equation of material balance for an infinitesimally small volume of the macrofracture.

It has the form (see Chap. 1)

$$\frac{\partial C_{\mathrm{I}}}{\partial t} + u\,\frac{\partial C_{\mathrm{I}}}{\partial x} - D\,\frac{\partial^2 C_{\mathrm{I}}}{\partial x^2} - D\,\frac{\partial^2 C_{\mathrm{I}}}{\partial y^2} = 0, \tag{7.1}$$

where C_{I} is the concentration of dissolved substance in the macrofracture space, and D is the coefficient of diffusion in the macrofracture space (it coincides with the coefficient of molecular diffusion).

Equation (7.1) shows that a change in concentration in the volume of mobile phase (liquid or gaseous) in the macrofracture is due to transfer of substance by filtration flow ($u\,\partial C_{\mathrm{I}}/\partial x$) and also by diffusion of the substance along the fracture ($D\,\partial^2 C_{\mathrm{I}}/\partial x^2$) and perpendicular to the walls ($D\,\partial^2 C_{\mathrm{I}}/\partial y^2$). The equation of material balance for an infinitesimally small volume of country rock is written in the form (Chap. 1)

$$\frac{\partial C_{\mathrm{II}}}{\partial t} + \frac{\partial q}{\partial t} - \overline{D}\,\frac{\partial^2 C_{\mathrm{II}}}{\partial x^2} - \overline{D}\,\frac{\partial^2 C_{\mathrm{II}}}{\partial y^2} = 0, \tag{7.2}$$

where C_{II} is the concentration of dissolved substance in the microfractures, q is the concentration of substance transferred in the immobile phase (rock) as a result of adsorption, ion exchange, or chemical reaction, and D is the diffusion coefficient in the microfractures, which in the general case differs from D.

Equation (7.2) accounts for the change in concentration of solution in the microfracture space ($\partial C_{\mathrm{II}}/\partial t$) due to diffusion and adsorption or chemical reactions ($\partial q/\partial t$).

The third equation describing the process is the kinetic equation of the process of physicochemical interaction between solution and country rock, which describes the interaction between substance and country rock in time at any fixed point. Kinetic equations of the physicochemical processes of adsorption, ion exchange, and chemical reactions were examined above. However, we might write the kinetic equation in the general form with details of the process of interaction between solution and rock. In the general case, the rate of the process is determined by the concentration of the substance in solution and in the rock phase, the rate constants K_i of the chemical reactions between solution and the medium, the diffusion coefficients, and the flow rate. Mathematically this dependence may be written implicitly:

$$\frac{\partial q}{\partial t} = \varphi\,(C,\,q,\,K_i,\,u,\,D,\,\overline{D}). \tag{7.3}$$

The initial and boundary conditions for the investigated process may be written in the form

$$t > 0, \quad x = 0, \qquad\qquad C_{\mathrm{I}} = C_0 \text{ (condition of symmetry)}, \tag{7.4}$$

$$y \to \infty, \qquad\qquad C_{\mathrm{II}} = 0 \text{ (condition of semiinfinite medium)}, \tag{7.5}$$

$$y = \frac{l}{2}, \qquad\qquad C_{\mathrm{I}} = C_{\mathrm{II}} \text{ (condition of equality of concentrations} \tag{7.6}$$
$$\text{at the boundary with the wall)},$$

$$D\left(\frac{\partial C_{\mathrm{I}}}{\partial y}\right)_{y=l/2} = \overline{D}\left(\frac{\partial C_{\mathrm{II}}}{\partial y}\right)_{y=l/2} \text{ (condition of equality of flow movements} \tag{7.7}$$
$$\text{at the boundary with the wall)},$$

$$y = 0, \qquad\qquad\qquad \frac{\partial C_I}{\partial y} = 0 \quad \text{(condition of symmetry)} \qquad\qquad (7.8)$$

$$t = 0, \qquad x > 0, \qquad\qquad C_I = 0, C_{II} = 0 \qquad\qquad\qquad (7.9)$$

The solution of the problem we have formulated depends on the form of the kinetic equation (7.3). However, even for the simplest kinetic equations it is difficult to obtain a precise solution of the system of equations (7.1)–(7.3) with conditions (7.4)–(7.9). It is necessary to introduce some simplifying assumptions. We shall assume that the diffusion coefficient $\overline{D}$ of the substance in the microfractures is so low that the substance penetrates but an inappreciable distance into the walls.

With this approximation, it is convenient to change from concentrations $C_I(x, y, t)$ and $q(x, y, t)$ to concentrations $C(x, t)$ and $q(x, t)$, averaged for the section perpendicular to the x axis, and to write the equation of material balance for the x section of the fractures. It has the form

$$\frac{\partial C}{\partial t} + u \frac{\partial C}{\partial x} + \frac{\partial q}{\partial t} - D \frac{\partial^2 C}{\partial x^2} = 0. \qquad\qquad (7.10)$$

Equation (7.10) shows that in the x section any change in average concentration of substance of the mobile phase $(\partial C/\partial t)$ is due to filtration of the substance $(u \partial C/\partial x)$, diffusion along the fracture $(D \partial^2 C/\partial x^2)$, and interaction with the walls $(\partial q/\partial t)$. In this case, the system of initial and boundary conditions takes on the form

$$t = 0, \quad x > 0, \qquad\qquad C = 0, \qquad\qquad\qquad\qquad (7.11)$$

$$q = 0, \qquad\qquad\qquad\qquad (7.12)$$

$$t > 0, \quad x = 0, \qquad\qquad C = C_0. \qquad\qquad\qquad\qquad (7.13)$$

The system of differential equations (7.3) and (7.10) with conditions (7.11)–(7.13) describes the model we have adopted for the formation of primary geochemical aureoles of ore deposits and gas deposits. The system is analogous to equations describing adsorption dynamics and chromatography, and also chemical reactions taking place in a current. Solution of these equations were obtained in Chap. 6. We shall use them in our further discussions.

Below we shall examine separately solutions of the formulated problem for interaction processes between the dissolved substance and the country rocks: 1) adsorption and ion exchange, 2) irreversible chemical reaction of the second order, 3) reversible chemical reaction of the first order, and 4) irreversible chemical reaction taking place in the diffusion region.

The Formation of Primary Geochemical Aureoles in the Presence of Adsorption and Ion Exchange. The formation of primary geochemical aureoles of ore deposits and gas deposits with consideration of adsorption and ion exchange without chemical reactions represents only a particular case. The formation of geochemical aureoles of any type are generally accompanied by adsorption and ion exchange between the substance and the country rock, however. Therefore, an examination of the indicated problem is of definite interest.

In order that a molecule of the dissolved substance be adsorbed by the rock, it is necessary to transfer it from the bulk solution to the surface of the rock. When adsorption (ion exchange) takes place on the walls of the macrofracture, the transfer of the substance is effected by diffusion from a current or by convective diffusion. If adsorption (ion exchange) takes place on the walls of the microfractures, the transfer of the substance from the bulk solution takes place successively by convective diffusion in the macrofracture and by diffusion along the microfractures. By analogy with the terminology used in Chap. 6, we shall agree to call the transfer of substance from the bulk solution to the wall of the macrofracture external diffusion, and transfer in the microfractures we shall call internal diffusion.

Numerous experiments on the kinetics of adsorption and ion exchange (see Chap. 4) have shown that diffusion normally takes place much more slowly than adsorption or ion exchange. It may therefore be assumed that on the surface of the phase boundary (rock–solution) equilibrium between adsorbed substance and rock exists at any instant of time. Consequently, the concentration of adsorbed substance q is related to the concentration $\overline{C}$ of the solution adjacent to the rock surface by the adsorption (ion-exchange) isotherm:

$$\overline{C} = f(q). \tag{7.14}$$

The rate of adsorption (ion exchange) in this case is determined by diffusion. If the rate of adsorption (ion exchange) is determined by the diffusion of substance to the walls of the macrofracture, we shall state that the formation of the primary aureole takes place in the external-diffusion region. The equation of external-diffusion kinetics of adsorption (ion-exchange) has the form (see Section 35)

$$\frac{\partial q}{\partial t} = \gamma_1 [C - f(q)], \tag{7.15}$$

where γ_1 is the kinetic coefficient (rate constant) of external diffusion, depending on the diffusion coefficient D and the flow rate u.

If the rate of adsorption (ion exchange) is determined by the diffusion of substance in the microfractures, we shall state that the process takes place in the internal-diffusion region. The equation of internal-diffusion kinetics of adsorption (ion exchange) may be written in the following form (see § 36):

$$\frac{\partial q}{\partial t} = \gamma_2 [C - f(q)], \tag{7.16}$$

where γ_2 is the kinetic coefficient of internal diffusion, depending on $\overline{D}$.

If the rates of external and internal diffusion are similar, the formation of a primary aureole takes place in the mixed diffusion region. The kinetic equation for simultaneous consideration of external and internal diffusion has the form (see § 37)

$$\frac{\partial q}{\partial t} = \gamma [C - f(q)], \tag{7.17}$$

where the rate constant of the process is equal to

$$\frac{1}{\gamma} = \frac{1}{\gamma_1} + \frac{1}{\gamma_2}. \tag{7.18}$$

From the general theory of geochemical migration, it follows that the system of equations (7.10) and (7.15), (7.10) and (7.16), and (7.10) and (7.17) with conditions (7.11)-(7.13) describes the formation of a primary aureole in the external-diffusion, internal-diffusion, and mixed regions respectively.

Let us consider the solution of these systems for the case in which a solution of low concentration flows along the macrofracture, a situation that apparently corresponds to natural conditions in the formation of geochemical aureoles of a number of elements. The adsorption (ion-exchange) isotherm will then become a linear function of the concentration:

$$\bar{C} = f(q) = \frac{q}{K} , \tag{7.19}$$

where K is the adsorption coefficient (or distribution in the case of ion exchange).

An asymptotic solution (for large time values) of the system of equations (7.10), (7.17), and (7.19) with conditions (7.11)-(7.13) was obtained in Chap. 6. It has the form

$$C(x, t) = \frac{C_0}{2} \left[1 - \mathrm{erf} \left(\frac{x - \frac{u}{1+K} t}{2 \sqrt{\frac{D_{lo}}{1+K} t}} \right) \right], \tag{7.20}$$

$$q(x, t) = \frac{K C_0}{2} \left[1 - \mathrm{erf} \left(\frac{x - \frac{u}{1+K} t}{2 \sqrt{\frac{D_{lo}}{1+K} t}} \right) \right], \tag{7.21}$$

where

$$\mathrm{erf}\, z = \frac{2}{\sqrt{\pi}} \int_0^z e^{-y^2} dy .$$

The value D_{lo} is called the effective coefficient of longitudinal diffusion and is equal to (see §45)

$$D_{lo} = D + \frac{K^2 u^2}{(1+K)^2} \left(\frac{1}{\gamma_1} + \frac{1}{\gamma_2} \right), \tag{7.22}$$

where D is the diffusion coefficient in the bulk solution.

From Eq. (7.20) it follows that when filtration of a substance takes place along the fracture a front that spreads with time is formed, i.e., the distance at which the concentration changes (continuously) from 0 to C_0 increases with time. The front becomes longer the larger the value of D_{lo}. The first term in Eq. (7.22) here describes the diffuseness due to the diffusion of the substance along the fracture; the second and third describe external and internal diffusion respectively. Equations (7.20)-(7.22) describe the process in its most general form, including as particular cases the solutions of the problem of forming a geochemical aureole in the internal-diffusion region [when the second term in Eq. (7.22) is small and may be neglected] and the problem of forming an aureole in the external-diffusion region (when the third term may be neglected). The distribution of adsorbed substance along the fracture [Eq. (7.21)] is like that shown in Fig. 27.

We shall compute the distance of geochemical migration of the substance, meaning the greatest distance at which the adsorbed substance may still be detected by quantitative analysis.

Without making any great error, when q/q_0 is small, we may write in place of Eq. (7.21) the following:

$$q\,(x,\,t) = \frac{q_0}{2}\left[1 - \left(\frac{x - \dfrac{u\,t}{1+K}}{\sqrt{\pi\,\dfrac{D_{lo}}{1+K}\,t}}\right)\right],\qquad (7.23)$$

where

$$q_0 = KC_0.$$

Substituting in Eq. (7.23) the minimal concentration q_{min} in place of $q\,(x,\,t)$ (q_{min} being the smallest concentration that may still be determined) and x_{max}, the distance of migration in place of x, we find

$$x_{max} = \frac{u}{1+K}\,\tau + \left(1 - \frac{2q_{min}}{q_0}\right)\sqrt{\pi\,\frac{D_{lo}}{1+K}\,\tau},\qquad (7.24)$$

where τ is the time required for formation of the ore body. If $q_{min} = \alpha q_0$, then

$$x_{max} = \frac{u}{1+K}\,\tau + \sqrt{\pi\,\frac{D_{lo}}{1+K}\,\tau}\,(1-2\alpha),\qquad (7.25)$$

which coincides with Eq. (6.70). By using Eqs. (7.24) and (7.25), it is possible to compute the greatest distance at which we may detect substance of the ore body if we know the values of D_{lo}, K, u, and τ.

The Formation of Primary Geochemical Aureoles with Considera-tion of Chemical Reactions. The formation of primary geochemical aureoles of ore deposits and gas deposits with consideration of chemical reactions represents the most general case, a detailed examination of which is of considerable interest, particularly for the theory of ore formation.

Chemical reaction between a solution and any component of the rock takes place at the phase boundary (rock−solution) and is a heterogeneous process. If the rate at which the sub-stance diffused from the bulk solution to the rock is greater than the reaction rate, then the rate of interaction between solution and rock is equal to the rate of chemical reaction. The formation of a geochemical aureole takes place in this case in the kinetic region. On the other hand, if the supply of substance to the rock is slower than the reaction itself, the process then takes place in the diffusion region (see Chap. 4).

The problem of forming primary geochemical aureoles, with consideration of chemical reactions, may be simplified if we assume that because of a low diffusion rate the amount of substance transferred by diffusion flow is generally considerably less than the amount trans-ferred by filtration flow. Then, in Eq. (7.10) member $(D\,\partial^2 C/\partial x^2)$ is negligible as compared with $(u\,\partial C/\partial x)$, and the equation may be written

$$\frac{\partial C}{\partial t} + u\,\frac{\partial C}{\partial x} + \frac{\partial q}{\partial t} = 0.\qquad (7.26)$$

The problem of forming a primary aureole is described by a system of an equation of material balance (7.26) and a kinetic equation of chemical reaction (see Chap. 4) with conditions (7.11)-(7.13). We shall consider a solution of the formulated problem for different chemical reactions between the substance and the medium. In this, it will be assumed that the rate of heterogeneous chemical reaction in the kinetic range may be described by equations of formal kinetics.

1. Irreversible Reaction of the Second Order.

Let us examine the chemical reaction (5.44). In accordance with the basic postulate of chemical kinetics [6], the equation of reaction rate (5.44) for the case in which diffusion does not retard the process has the form of (5.45).

A solution of the system of equations (7.26) and (5.45) with conditions (7.11)–(7.13) [7] was presented in Chap. 6: Eqs. (6.93) and (6.94). We shall investigate Eq. (6.94), which gives the distribution of ore substance along the fracture.

Let the concentration of rock components with which the solution interacts be low ($q_0 \ll C_0$, $q_0 \to 0$). It is easy to see that Eq. (6.94) then takes on the form

$$q(x, t) = q_0 \left[1 - e^{-K'C_0 \left(t - \frac{x}{u} \right)} \right]. \tag{7.27}$$

Equation (7.27) coincides with Eq. (6.83) of geochemical migration due to filtration in a slightly adsorbent medium with the condition

$$\gamma = K'C_0. \tag{7.28}$$

The distribution of ore substance along the filtration path of the solution (x axis) is shown in Fig. 36 for $q_0 < C_0$ ($q_0 = 0.2C_0$).

If the concentration of the active component of the rock is large ($q_0 \gg C_0$, $C_0 \to 0$), Eq. (6.94) by a simple transformation is brought approximately to the form

$$q(x, t) = K'q_0C_0 \left(t - \frac{x}{u} \right) e^{-K'q_0 \frac{x}{u}}. \tag{7.29}$$

The solution of Eq. (7.29) coincides with the solution (6.100) of the problem of one-dimensional filtration with consideration of first-order irreversible reaction between dissolved substance and country rock if we set

$$\gamma = K'q_0. \tag{7.30}$$

The distribution of ore substance along the fracture is shown in Fig. 37, where $q_0 > C_0$ ($q_0 = 5C_0$).

In the general case, when the values of q_0 and C_0 are similar, it is necessary to use Eq. (6.94) to plot the concentration profiles. The distribution curves for concentrations $q(x, t)$ for the particular case $q_0 = C_0$ are shown in Fig. 38.

From Figs. 36–38 we see that diffuse fronts of the reacting substances are formed in the fracture (concentrations in the front range continuously from zero to C_0 or q_0).

From the expression (6.94) it follows that for time t the aureole is distributed from the ore body to a distance $x_{max} = ut$. The distance of migration may be computed, as above, by substituting the values q_{min} and x_{max} in Eq. (6.94) in place of $q(x, t)$ and by solving the resulting equation for x_{max}.

2. Reversible Reaction of the First Order.

Let the reaction (5.46) take place between the dissolved substance and minerals of the country rock. The equation of reaction

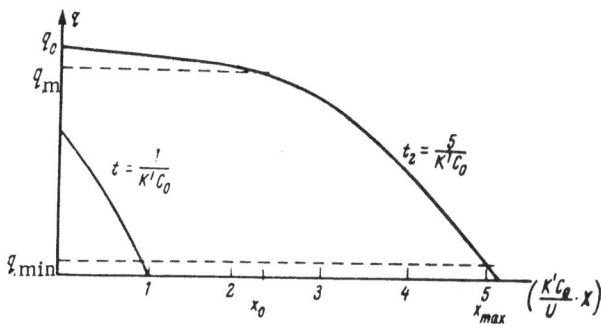

Fig. 36. Distribution of ore substance having formed as a result of irreversible second-order reaction in the solid phase at $q_0 = 0.2C_0$.

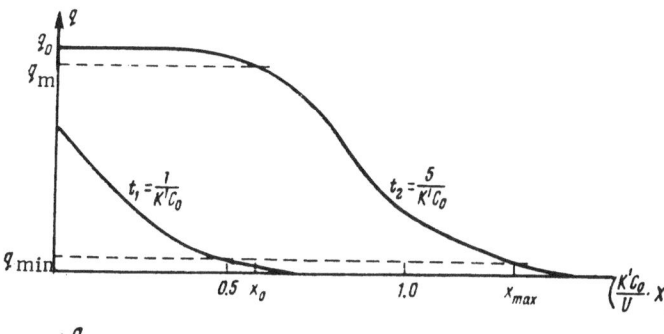

Fig. 37. Distribution of ore substance having formed as a result of irreversible second-order reaction in the solid phase at $q_0 = 5C_0$.

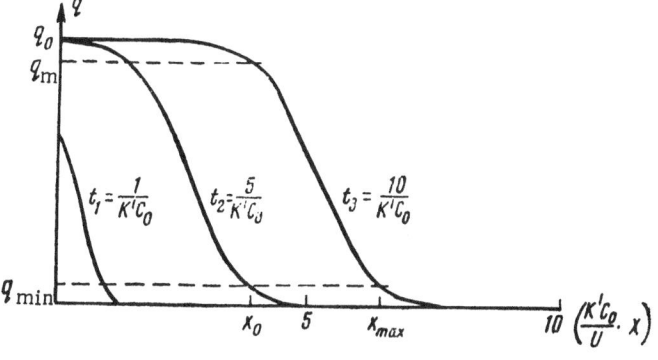

Fig. 38. Distribution of ore substance having formed as a result of irreversible second-order reaction at $q_0 = C_0$.

rate for this has the form (5.47). This equation gives the rate of first-order reversible reaction similar to Eqs. (7.15)–(7.17) of the diffusion kinetics of adsorption for a linear isotherm (7.19). In this case the primary geochemical aureole is described by Eqs. (7.20)–(7.23), if we set

$$D_{lo} = \frac{K_c^2}{(1+K_c)^2} \frac{u^2}{K_1}, \qquad (7.31)$$

where $K_c = K_1/K_2$ are the equilibrium constants of reaction (5.46). The distribution of substance B in the solid phase has a form similar to the distribution of substance in Fig. 27. The expression for distance of geochemical migration is found in accordance with Eq. (7.24), and it has the form

$$x_{max} = \frac{u}{1+K_c}\tau + \left(1 - \frac{2q_{min}}{q_0}\right)\frac{K_c u}{1+K_c}\sqrt{\pi\frac{\tau}{K_1}}, \qquad (7.32)$$

where τ is the time required for formation of the ore body. If $q_{min} = \alpha q_0$, then

$$x_{max} = \frac{u}{1+K_c}\tau + \frac{K_c u}{1+K_c}\sqrt{\pi\frac{\tau}{K_1}}\,(1-2\alpha). \qquad (7.33)$$

By using Eq. (7.33) we may compute the distance of geochemical migration for any time if we know the values of u, K, K_1, and τ.

3. <u>Irreversible Reaction Taking Place in the Diffusion Region.</u> Let us consider reaction (5.44). If it takes place in the diffusion region, the equation of rate in accordance with Fick's first law of diffusion may be written by means of Eq. (5.61). The system of equations (7.26) and (5.61) describes the formation of a primary geochemical aureole for irreversible reaction between solution and rock in the diffusion region of kinetics. A solution of this system for conditions (7.11)–(7.13) was obtained in Chap. 6. The distribution of substance in the immobile phase in described by Eq. (6.100).

Equation (6.100) also describes the distribution of substance in the solid phase if first-order irreversible reaction takes place between solution and rock. Actually, the equation of rate (6.32) of the first-order irreversible reaction is similar to Eq. (5.61) of the diffusion kinetics of irreversible second-order reaction if we set $\gamma = K$ (γ and K being the rate constants of diffusion and irreversible first-order reaction respectively). Irreversible reaction takes place as a first-order reaction if the rock with which the solution reacts is of monomineralic composition.

The distance of geochemical migration x_{max} for the time t may be determined from the following equation, written in implicit form:

$$\frac{q_{min}}{C_0} = \gamma \left(\tau - \frac{x_{max}}{u} \right) e^{-\frac{\gamma}{u} x_{max}} . \qquad (7.34)$$

At the limit, when $q_{min} \ll C_0$, $q_{min} \to 0$, we find that the aureole is spread from the ore body to a distance $x_{max} = u\tau$.

Figure 39 shows the distribution of substance in the solid phase computed by Eq. (6.100). A similar equation was obtained by Dubov [4].

When the reaction between mobile and immobile phases is reversible, the equation of diffusion kinetics of the reaction coincides with Eq. (7.15) of adsorption kinetics for a linear isotherm (7.19), if by K we mean the equilibrium constant K_c of reaction (5.46). Therefore, to describe the formation of a primary aureole, we may in this case use the solutions of Eqs. (7.20)–(7.25), the properties of which were discussed above.

<u>The Formation of a Primary Geochemical Aureole with Filtration of Mixtures.</u> The formation of primary geochemical aureoles is normally effected by filtration of mineral-forming solutions of very complex mixtures. The features of aureole formation by filtration of complex solutions is of real interest from the viewpoint of solving a number of basic questions on the theory of ore formation. If a mixture of substances that react with the walls filters along the microfracture, the problem is substantially complicated. However, in a number of cases it is possible to obtain a solution of the problem represented by filtration of mixtures.

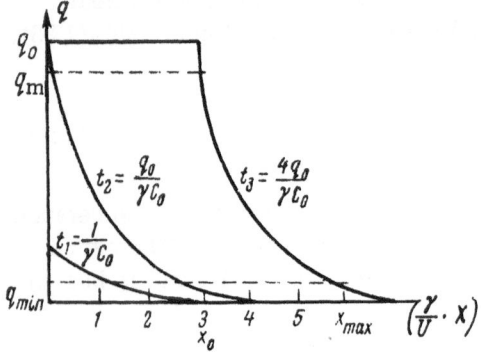

Fig. 39. Distribution along the path of solution filtration of ore substance having formed as a result of first-order irreversible reaction.

Let each of the components of the solution be adsorbed by or exchange ions with the rocks. If the adsorption (ion-exchange) isotherm is linear, the transfer of each substance to the rock takes place independently of the others. Let the coefficient of adsorption of i substance be K_i (i = 1, 2, ..., n the number of components of the mixture). Then the distribution of the i substance in the medium has the form (7.20)-(7.23) if we replace K by the adsorption coefficients K_i of the individual substances of the mixture. Thus, we shall have a set of solutions, differing from each other by the values of K_i (i = 1, 2, ..., n). By investigating such solutions (Chap. 6) it is possible to establish the fact that, beginning at some point in time, differential flow of the components of the solution occurs: each substance will move at its characteristic rate v_i, equal to

$$v_i = \frac{u}{1 + K_i} \quad (i = 1, 2, \ldots, n). \tag{7.35}$$

From the above expression, it follows that the better a substance is adsorbed the lower the rate of its movement along the fracture. Schematically the distribution of adsorbed components of the solution along the fracture is similar to the distribution shown in Fig. 30. From Eq. (7.31) and Fig. 30 it is seen that a zonal distribution of adsorbed components on the walls of the macrofracture develops. In this picture, the better a substance is adsorbed, the nearer it is found to the ore body.

If the components of the mixture enter into chemical reactions with the country rock, then, as a first approximation, it may be assumed that each component reacts independently of the presence of the others. In this case, the distribution of the i component (i = 1, 2, ..., n) in the medium may be described by any of the above-investigated equations (depending on the nature of the reaction) if the individual rate constants and diffusion coefficients of a substance are substituted for the rate constant and diffusion coefficient of the i component of the solution (i = 1, 2, ..., n). Thus, as in the case of adsorption, we shall have a set of n solutions, differing from each other by the rate constants of the reactions. This also may serve as an explanation of the differential movement of the components in the solution, and, as a consequence, of the zonal distribution resulting from reactions of the substances along the macrofracture. The results obtained are supported by the theoretical considerations of Korzhinskii in his empirical principles of differential mobility of elements in the earth's crust [8].

The evolved theory describes the formation of hydrothermal deposits by chemical reactions between a solution and the country rock. An ore body is formed when the concentration q of the ore substance formed by reaction exceeds the marginal concentration q_m. When the given conditions are satisfied, the equations obtained above, (6.94), (6.100), (7.21), (7.27), and (7.29) give the concentration distribution in the ore body along the filtration path of the ore-bearing solution (x axis). The formation of hydrothermal deposits by crystallization during changes in thermodynamic conditions (pressure, temperature) of filtration of dissolved substances requires no special examination; it is described by Eq. (6.100) if by γ we mean the rate constant of crystallization. This is explained by the fact that, in the simplest case, crystallization may be formally considered a first-order irreversible reaction [9].

Equations (6.94), (6.100), (7.21)-(7.25), (7.27), (7.29), (7.33), and (7.34) may be used for quantitative interpretation of the results of a geochemical survey. It must be emphasized that the boundary between an "ore concentration" and a primary aureole is tentative [10]. If q_m is the marginal concentration of ore substance on Figs. 36-39, then x_0 is the extent of the ore body and $x_{max} - x_0$ is the extent of the primary aureole. Depending on the value of q_m, the extent of the ore body and its aureole will vary. From the equations we have obtained it follows that, in order to describe a geochemical aureole, it is necessary to know the duration of the ore-forming process, the equilibrium constants, the rates of adsorption, diffusion, and chemical reaction, and other parameters. In individual cases, these may be found experimentally.

This course is laborious, however, and is seldom followed in practice. It is therefore, advisable, in describing the aureole, to use different dimensional and dimensionless parameters [5, 10], which may be found for several concentration points in the aureole. For such parameters we may choose, in particular, the values $\left(\frac{u}{1-K}\tau\right)$, $\left(\frac{D}{1+K}\right)$, $\gamma\tau$, $\left(\frac{K'q_0}{u}\right)$, and others. The question of the suitability of any particular parameter of the aureole requires supplementary investigation.

To determine the parameters of the aureole, it is convenient to plot the distribution curves of anomalous concentrations not in the coordinates (q, x) as in Figs. 36–39, but in others that permit rectification of the curves. For example, the concentration distribution for irreversible first-order reaction [Eq. (7.29)] may be conveniently plotted in the coordinates $(\log q, x)$. The slope gives the value of the aureole parameter $|K'q_0/u|$. What we have said refers to both primary and secondary aureoles, the formation of which are discussed below.

In conclusion, we may note that, as seen from Figs. 36–39 [and perhaps shown by investigation of Eqs. (6.94), (6.100), (7.27), and (7.29)], three most characteristic shapes of concentration curves of primary aureoles may be encountered, depending on the condition of formation: curves having a point of inflection (Figs. 37–38), curves having no inflection point and being convex upward (Fig. 36), and curves without inflection point, convex downward (Fig. 39).

§ 54. The Formation of Deposits by Reaction between Components of Hydrothermal Solutions

The formation of hydrothermal deposits may take place by interaction between components of different solutions that become mixed by filtration and diffusion. The important roles of these processes in ore formation have been especially emphasized by Pospelov [11]. This investigator and Kaushanskaya [12, 13] experimentally observed the formation of minerals as a result of chemical reaction between diffusing ions.

We shall examine the formation of a deposit by reaction between components of hydrothermal solutions during counter diffusion. We shall adopt the following model of the process. Let solutions A and B with constant concentrations C_{01} and C_{02} move by filtration along two parallel macrofractures separated from each other by a distance l. The surrounding rocks are porous or contain an isotropic network of microfractures, filled with aqueous solutions that are in equilibrium with the minerals of the rocks. Let the substances A and B not interact with the rocks but react irreversibly between themselves, giving rise to one or several insoluble compounds. Then, at the place the diffusion currents of A and B meet, insoluble reaction products precipitate out, forming new minerals.

The problem of counter diffusion of substances with reactions taking place has been solved by several investigators [14, 15], but they have not accounted for the kinetics of the process. Instead, they have used the law of mass action [16]. However, the investigated process is not an equilibrium process (in the thermodynamic sense [16]), and the use of the equilibrium constant is therefore not rigid.

We shall examine the case of second-order irreversible reaction between the diffusion substances, with the formation of relatively insoluble compounds. In the general case, when supersaturation is not zero, the reaction may be schematically written as

$$A + B - C_l + D + \ldots, \tag{7.36}$$

$$C_l \rightleftarrows C_{so} \tag{7.37}$$

where C_l and C_{so} are the low-solubility products of reaction C in the liquid and solid phases.

We shall set the x axis perpendicular to the fractures. We are required to determine the distribution law of concentration of the substance C in the country rocks at any point in time.

Equations of the material balance of substances A and B for an infinitesimally small volume of rock in the space between the fractures may be conveniently written

$$\frac{\partial C_1}{\partial t} + w = D_1 \frac{\partial^2 C_1}{\partial x^2},$$ (7.38)

$$\frac{\partial C_2}{\partial t} + w = D_2 \frac{\partial^2 C_2}{\partial x^2},$$ (7.39)

where C_1 and C_2 are the concentrations of substances A and B, D_1 and D_2 are the diffusion coefficients, and w is the rate of reaction (7.36).

The rate of reaction (7.36) is equal to

$$w = K_1 C_1 C_2,$$ (7.40)

where K_1 is the rate constant of the reaction.

Let us find the relation between the reaction rate w and the concentration q (g/cm^3) of the mineral that is formed. The rate of the crystallizing process (7.37) in the simplest case is proportional to the degree of supersaturation [9]:

$$\frac{\partial q}{\partial t} = K_2 (C_3 - C_s),$$ (7.41)

where C_3 is the concentration of the substance C in the liquid phase, C_s is the saturation concentration, and K_2 is the rate constant of crystallization.

The value of C_3 may be found from the equation showing change in the substance C_l with time:

$$\frac{\partial C_3}{\partial t} = w - K_2 (C_3 - C_s).$$ (7.42)

The rate constant of crystallization K_2 is proportional to the area of the surface of the precipitate, which in the general case increases with time. However, if one-dimensional growth of the crystals takes place, then K_2 does not depend on time. Below, for simplicity, we shall consider the rate constant of crystallization to be constant. This assumption has no substantial effect on the pattern of the investigated phenomena.

We shall assume that at the initial moment of time the substances A and B are absent from the country rock in the zone between the fractures $0 < x < l$. The initial and boundary conditions of the problem are written in the form

$$\begin{cases} x = 0, & t > 0, & C_1 = C_{01}, \\ x = l, & t > 0, & C_2 = C_{02}, \\ t = 0, & 0 < x < l, & C_1 = 0, \\ & & C_2 = 0, \\ & & C_3 = 0, \end{cases}$$ (7.43)

The system of differential equations (7.38)-(7.42) with conditions (7.43) describes the model we have adopted for the formation of hydrothermal deposits by irreversible second-order chemical reaction during counter diffusion of the substances. The system is not linear; its solution therefore presents considerable diffuculty.

We shall seek a solution of the problem for a particular case, in which the rate of reaction (7.36) is low:

$$w \ll \left| \frac{\partial C_1}{\partial t} \right|; \qquad w \ll \left| \frac{\partial C_2}{\partial t} \right|, \tag{7.44}$$

so that the change in concentrations C_1 and C_2 in the space $0 < x < l$ takes place chiefly because of diffusion. Furthermore, we shall consider the initial stage of the process when the front of diffusing substances does not reach the opposite fractures. The space $0 < x < l$ for each diffusing substance is then semiinfinite. The solution of Eqs. (7.38) and (7.39), considering the inequality (7.44) for conditions (7.43), is written in the form

$$C_1 = C_{01} \left[1 - \operatorname{erf} \left(\frac{x}{2\sqrt{D_1 t}} \right) \right], \tag{7.45}$$

$$C_2 = C_{02} \left[1 - \operatorname{erf} \left(\frac{l-x}{2\sqrt{D_2 t}} \right) \right], \tag{7.46}$$

where erf is the designation of the Gaussian integral. Taking into account that when C_1 and C_2 are small the Gaussian integral may be expanded into a series and approximately equated with the independent variable, we obtain by substituting from (7.45) and (7.46) in Eq. (7.40) the following equation for the rate of reaction (7.36):

$$w = K_1 C_{01} C_{02} \left(1 - \frac{x}{2\sqrt{D_1 t}} \right) \left(1 - \frac{l-x}{2\sqrt{D_2 t}} \right). \tag{7.47}$$

By substituting this value in Eq. (7.42) and integrating, we may find the value of C_3. By integrating Eq. (7.41) the desired value of $q(x, t)$ may be obtained. However, cumbersome mathematical expressions result from this procedure.

For simplicity let us consider two limiting cases.

C a s e 1. Let supersaturation be infinitely large; i.e., as a result of reaction (7.36) we obtain practically insoluble precipitate. Then

$$\frac{\partial q}{\partial t} = w. \tag{7.48}$$

By substituting in (7.48) the expression for w from Eq. (7.47) and integrating, we obtain

$$q = K_1 C_{01} C_{02} \left[t - \left(\frac{x}{\sqrt{D_1}} + \frac{l-x}{\sqrt{D_2}} \right) \sqrt{t} + \frac{x(l-x)}{2\sqrt{D_1 D_2}} \ln t \right] + \varphi(x), \tag{7.49}$$

where $\varphi(x)$ is the constant of integration, depending on x in the general case. To find $\varphi(x)$ we use the following method. The distance of migration of substances A and B is approximately equal to

$$x_{max_1} = 2\sqrt{D_1 t}, \tag{7.50}$$

$$x_{max_2} = l - 2\sqrt{D_2 t}. \tag{7.51}$$

Whence it is easy to find that the meeting of diffusion currents of substances A and B takes place at time τ, which is equal to

$$\tau \approx \frac{l^2}{4 \left(D_1 + D_2 + 2 \sqrt{D_1 D_2} \right)}. \tag{7.52}$$

It is clear than when $t = \tau$ (also when $t < \tau$) $q = 0$. Finding $\varphi(x)$ by means of this condition, we obtain the final solution

$$q = \begin{cases} K_1 C_{01} C_{02} \left[(t - \tau) - \left(\frac{x}{\sqrt{D_1}} + \frac{l-x}{\sqrt{D_2}} \right) \left(\sqrt{t} - \sqrt{\tau} \right) + \frac{x(l-x)}{2 \sqrt{D_1 D_2}} \ln \frac{t}{\tau} \right], & t \geqslant \tau, \\ \\ 0, & t \leqslant \tau. \end{cases} \tag{7.53}$$

For the particular case when the mobility of substance A is the same as that of B $(D_1 = D_2)$, the solution takes the form

$$q = \begin{cases} K_1 C_{01} C_{02} \left[\left(t - \frac{l^2}{16D} \right) - \frac{l}{\sqrt{D}} \left(\sqrt{t} - \frac{l}{4\sqrt{D}} \right) + \frac{x(l-x)}{2D} \ln \frac{16Dt}{l^2} \right], \\ \hfill t \geqslant \frac{l^2}{16D}, \\ \\ 0, \hfill t \leqslant \frac{l^2}{16D}. \end{cases} \tag{7.54}$$

Differentiating Eq. (7.53) according to x and setting the resulting expression equal to zero, we find the point x^* in the medium where the concentration q is a maximum:

$$x^* = \frac{l}{2} + \frac{\sqrt{t} - \sqrt{\tau}}{\ln \frac{t}{\tau}} \left(\sqrt{D_1} - \sqrt{D_2} \right). \tag{7.55}$$

On the basis of Eqs. (7.53)-(7.55) we may make the following conclusions. When the mobilities of substances A and B are identical $(D_1 = D_2)$, the maximum concentration of q is found half way between the fractures $(x^* = l/2)$; it increases with time but does not change its position. When $D_1 \neq D_2$, the maximum is formed nearer the fracture along which the less mobile component migrates, and with time the maximum shifts closer to this fracture. This is shown schematically in Figs. 40 and 41.

C a s e   2 .   Let the rate of reaction (7.36) be considerably greater than the rate of crystallization (7.37) (great supersaturation). Then, in place of Eq. (7.42), we will have

$$\frac{\partial C_3}{\partial t} = w. \tag{7.56}$$

Substituting the value from (7.47) in Eq. (7.56) and integrating, we obtain an expression for $C_3(x, t)$ that coincides with Eq. (7.53) for $q(x, t)$. By substituting $C_3(x, t)$ in (7.41) and integrating the latter for conditions $t = \tau$, $q = 0$, we obtain for $C_3 > C_s$ the following distribution of substance C in the precipitate:

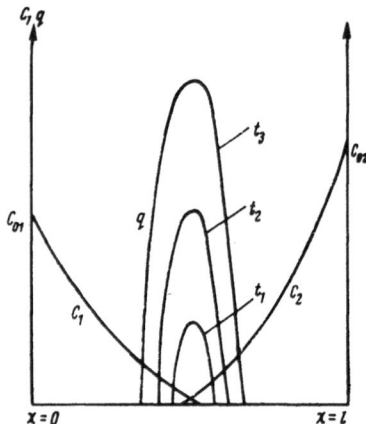

Fig. 40. Concentration curves during counterdiffusion of reacting substances for the case $D_1 = D_2$.

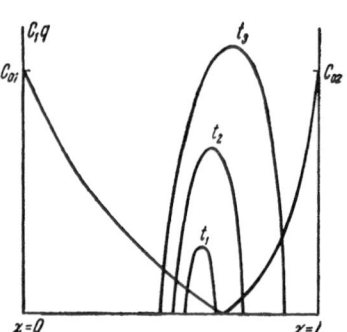

Fig. 41. Concentration curves during counterdiffusion of reacting substances for the case $D_1 \neq D_2$.

$$q = \begin{cases} K_1 K_2 C_{01} C_{02} \left[\left(\dfrac{t^2}{2} - t\tau + \dfrac{\tau^2}{2} \right) - \left(\dfrac{x}{\sqrt{D_1}} + \dfrac{l-x}{\sqrt{D_2}} \right) \times \\ \times \left(\dfrac{2}{3} t^{3/2} - t\sqrt{\tau} + \dfrac{\tau^{3/2}}{3} \right) + \dfrac{x(l-x)}{2\sqrt{D_1 D_2}} \left(t \ln \dfrac{t}{\tau} - t + \tau \right), \quad t > \tau, \\ 0, \qquad\qquad\qquad\qquad\qquad\qquad\qquad\qquad\qquad\quad t \leqslant \tau. \end{cases}$$

(7.57)

The solutions we have obtained are qualitatively confirmed by experimental results obtained by Pospelov and Kaushanskaya in model studies on the processes of mineral formation by counter diffusion of ions [12, 13]. In their studies they normally observed the following principal stages in the process of counterdiffusion of ions: the movement of precipitate toward the less mobile ion, interruption of the zone and growth of a mineral wall, further movement of the zone of precipitate and formation of a second wall, and so on. The solutions obtained above, as one should expect, describe only the initial stage of movement of the zone of precipitate, since the process of counterdiffusion is considered only for short times. Interruption and the formation of rhythmic zones of precipitate (Liesegang bands) must be predicted by a more complex theory, taking into account also the diffusion permeability of the medium as a result of formation of the precipitate.

§ 55. The Theory of Forming Secondary Geochemical Aureoles of Ore and Gas Deposits

The laws of the formation of secondary geochemical aureoles depend on the phase — solid, liquid, or gaseous — that is represented by the migrating substances of the deposit, and also on the depth at which the ore body occurs. It is advisable to consider separately a number of specific cases of the formation of secondary geochemical aureoles.

1. **Dissemination Aureoles of Ore Bodies Cropping out at the Surface.** The problem of forming a mechanical dissemination aureole was considered by Solovov [1]. On the basis of general considerations of probability theory, Solovov properly pointed out that a dissemination aureole from an infinitesimally thin deposit must be defined by the normal distribution law of Gauss. However, quantitative examination of the problem of forming a dissemination aureole above an infinitesimally thin deposit is not strictly possible,

as the author himself noted. The law he introduced to describe the mobility of particles of the deposit and the concept of "fictive velocity" of the particle movement are without foundation. Below we consider the problem on the basis of successive independent tests in the theory of probability [17].

Let us consider the simple case of weathering of substance of an infinite vertical vein, exposed at the surface. The x axis will be set perpendicular to the ground surface, the y and z axes in the ground surface. The coordinates of the vein are $0 < x < \infty$, $-\infty < z < +\infty$, $-\frac{1}{2}\Delta y \leqslant y \leqslant +\frac{1}{2}\Delta y$, where Δy = const is the thickness of the vein. We shall assume that the value of Δy is small, so that the vein may be considered as infinitesimally thin. We shall use Q to indicate the amount of substance arriving at a square centimeter of section perpendicular to x and shall postulate that at t = 0 the concentration of substance in the surrounding medium is zero. A great number of external forces, determining the mobility of the particles, act on each particle during weathering of the deposit. Let a particle be displaced a distance δy as a result of a single act of any force upon the particle. We shall assume that δy = const approximately and does not depend on the nature of the acting forces. If the direction along which the forces act are not preferential, the probabilities of displacement of the particle to distances $|+\delta y|$ and $|-\delta y|$ are identical and are equal to 0.5. Let us assume that the forces act independently. Then each displacement of the particle will be independent of the preceding displacement, and we may use the classic scheme of successive independent tests in the theory of probability to describe the movement of the particles. In accordance with this scheme, the probability that particles are displaced to a distance $y = m \cdot \delta y$ $(m = 1, 2, 3, \ldots)$ after they have been acted on n times [7], is equal to

$$\begin{cases} P = \dfrac{4n!}{(m+n)!\,(n-m)!}\left(\dfrac{1}{2}\right)^n, & (-n \leqslant m \leqslant +n), \\ 0, & |m| > n \ \ (n = 1, 2, 3 \ldots). \end{cases} \tag{7.58}$$

When $n \gg 1$, by using the Moivre-Laplace limit theorem, we obtain

$$P = \frac{1}{\sqrt{2\pi n}}\, e^{-\frac{m^2}{2n}}. \tag{7.59}$$

If the number of particles in the deposit is large, the value of P may then be considered as the fraction of particles displaced to a distance $y = m \cdot \delta y$. The concentration in the section x = 0 at the initial moment is equal to $Q/\Delta y$. If we set $\Delta y = \delta y$. Eq. (7.59) may then be written in the coordinates y and t in the following form:

$$q\,(x,t) = \frac{\dfrac{Q}{\Delta y}}{\sqrt{2\pi\dfrac{t}{\Delta t}}}\, e^{-\left(\frac{y}{\delta y}\right)^2 / 2\frac{t}{\Delta t}} = \frac{Q}{2\sqrt{\pi D^* t}}\, e^{-\frac{y^2}{4D^* t}}, \tag{7.60}$$

where q is the concentration of diffusing substance, Δt is the time between two successive displacements of a particle (we assume that Δt = const), and

$$D^* = \frac{(\delta y)^2}{2\Delta t}, \tag{7.61}$$

where D^* is a value analogous to the coefficient of diffusion.

Actually, expression (7.61) is similar to the Einstein equation relating the average displacement of a particle δy after time Δt, caused by thermal movement, to the diffusion coefficient. But Eq. (7.60) agrees with the distribution (2.40) for a substance diffusing from an infinitesimally thin layer (see Fig. 2). We shall call D^* the coefficient of mechanical dissemination. The movement of particles during weathering may with certain allowances be considered a quasi-diffusion process. The numerical value of the coefficient of mechanical dissemination is determined by the active forces of weathering, differing in character and in the nature of their actions, by virtue of which they are difficult to account for quantitatively. Theoretical calculation of D^* is therefore a problem of great difficulty.

Along with the disintegration and mechanical dissemination of substance of the ore deposit, diffusion and other processes of migration are at work. If we take diffusion into account, the dissemination aureole from an infinite vertical vein is then described by Eq. (7.60) if we replace D^* by the effective value

$$D_{\text{ef}} = D^* + D,$$

where D is the diffusion coefficient.

Many of the forces responsible for the mechanical breakup of a deposit act interdependently and with equal probability in any direction. Equation (7.60), therefore, as a first approximation, should properly describe the formation of mechanical dissemination aureoles from an infinite vertical vein (under conditions that unidirectional forces do not act on the deposit).

2. Secondary Dissemination Aureoles of Ore Bodies Covered by Unconsolidated Sediments. If an ore body is covered by rocks of the weathering zone and it occurs at considerable depth, then, as indicated above, the superposed dissemination aureole above the ore deposit forms by means of diffusion, filtration, and capillary raising of the ore substance. Since it is difficult to account quantitatively for all the processes at work in the formation of secondary dissemination aureoles, in seeking mathematical solutions of the problem we must neglect some of them. The most universal mechanism responsible for the formation of superposed dissemination aureoles may be assumed to be diffusion of dissolved substance through the pores of the unconsolidated rocks. As computations made by Antonov (see below) have shown, the distance of migration in geologic time may reach hundreds of meters by this process. We should keep in mind, however, that during diffusion, the dissolved substance is generally adsorbed by the rock or it reacts with the rock, and this greatly reduces the distance of migration.

Let us consider the problem of diffusion of a substance from an ore deposit through the pores of unconsolidated rocks under conditions that the rock is saturated with water. For simplicity, we shall assume that the substance is uniformly adsorbed by all the surrounding rock. The migration of substance in this case is described, as shown in Chap. 6, by the system of equation (6.1) of material balance for diffusing substance and Eq. (4.28) of the adsorption isotherm. We shall examine the diffusion of a substance of low concentration, which apparently corresponds with the natural environment in the formation of salt aureoles. In this case the adsorption isotherm becomes linear, and, in place of Eq. (4.28) it is necessary to use (3.11). By substituting from Eq. (3.11) in Eq. (6.1), we obtain the following equation describing the diffusion of a dissolved substance in an adsorbent medium:

$$\frac{\partial C}{\partial t} = \frac{D}{1+K} \Delta C, \tag{7.62}$$

where K is the coefficient of adsorption.

Let us consider the solutions of Eq. (7.62) for a number of cases that are of practical interest. In this consideration, we shall account for the following: Equation (7.62) is similar to the equation of nonstationary thermal conductivity; therefore, in examining diffusion, we may use the corresponding solutions of thermal-conductivity theory [18].

Diffusion from an Infinite Bed in the Presence of a Linear Diffusion Current. Let us assume for simplicity that the surface of the bed is parallel to the ground surface at a distance l from it. We shall assume also that these are plane surfaces, infinitely extended. The concentrations of diffusing substance at all points in planes perpendicular to the x axis (the directions of the axes remain as before) are identical, so that there occurs a linear diffusion flow of substance from the deposit along the x axis. Let the bed be large; then the concentration of substance going into solution from the ore deposit at the boundary x = l of the deposit does not change with the passage of time. If a nonvolatile substance diffuses, there is no flow of this substance into the atmosphere, and the boundary x = 0 (the erosional surface) is a reflecting surface. In accordance with this, the initial and boundary conditions for the investigated problem are written in the form

$$t = 0, \qquad 0 < x < l, \qquad C = 0,$$
$$t > 0, \qquad x = 0, \qquad \frac{\partial C}{\partial x} = 0, \tag{7.63}$$
$$t > 0, \qquad x = l, \qquad C = C_0,$$

where C_0 is the concentration of substance going into solution from the ore deposit.

Equation (7.62) with conditions (7.63) defines the distribution law of the substance in the medium at any point in time. A solution of Eq. (7.62) for conditions (7.63) may be obtained by the Laplace transform [19]. It has the form

$$\frac{C(x, t)}{C_0} = 1 + \frac{4}{\pi} \sum_{n=0}^{\infty} \frac{(-1)^{n+1}}{2n+1} \exp \left[-\frac{(2n+1)^2}{4l^2} \frac{\pi^2 Dt}{1+K} \right] \cos \left(\frac{2n+1}{2l} n\pi x \right). \tag{7.64}$$

Numerical calculations of the distribution of the diffusing substance in the pore solutions may be made by analyzing the convergence of the series (7.64) and summing the members of the series.

Let us determine the scale of diffusion for the investigated problem. The average concentration $\overline{C}$ of the substance in the pore solutions, as shown by computation, is equal to, on the basis of Eq. (7.64):

$$\frac{\overline{C}(t)}{C_0} = 1 - \frac{8}{\pi^2} \sum_{n=0}^{\infty} \frac{1}{(2n+1)^2} \exp \left[-\frac{(2n+1)^2}{l^2} \frac{\pi^2 Dt}{1+K} \right]. \tag{7.65}$$

Asymptotic investigation of equation (7.65) shows that for values $\overline{C}/C_0 < 0.5$ the distribution of diffusing substance in the medium may be represented approximately in the form

$$\frac{\overline{C}}{C_0} = \sqrt{\frac{\pi^2 Dt}{l^2 (1+K)}}. \tag{7.66}$$

From Eq. (7.66) it follows that for geologic intervals of time the average concentration $\overline{C}$ of the substance in pore solutions reaches large values, even when the covering rocks may be hundreds of meters thick. Thus, if we set t = 10^4 years, D = 10^{-6} cm^2/sec, l = 10^4 cm, and assume that adsorption is slight (K ≪ 1), it follows from expression (7.66) that $\overline{C}/C_0 \approx 0.2$. The distance of migration is determined by the greatest distance from the ore deposit at which the diffusing substance may be detected by analytical methods. It may be stated that C_0 agrees with the concentration of a saturated solution formed by solution in water of the product of chemical decomposition of the ore deposit. A concentration $\overline{C}$ of the diffusing substance equal to $0.2C_0$ may be measured without difficulty by analytical methods. Consequently, the distance of migration reaches hundreds of meters in the investigated example.

The computations made above are valid if we assume that the diffusion coefficient does not change with time and that adsorption is slight. If adsorption is large ($K \ll 1$), the distance of migration declines, in accordance with (7.66).

Diffusion from a Finite Bed. Let us examine diffusion from a deposit represented by a bed of thickness h, parallel to the ground surface, and infinitely extended in the directions of the y and z axes; i.e., the lateral extent of the bed is much greater than h. Since the thickness of the bed is finite, then, for any zone formed by planes parallel to the coordinates (x, 0, t) and (x, 0, z), the reserve of substance of the deposit, diffusing into it, is limited. Consequently, the concentration of substance entering into solution from the ore bed at the boundary of the deposit becomes dependent on time. The initial and boundary conditions for this problem are written in the form

$$
\begin{aligned}
t = 0, \qquad & 0 < x < l - h, \quad C = 0, \\
& l - h < x < l, \quad C = C_0, \\
t > 0, \qquad & x = 0, \qquad\quad \frac{\partial C}{\partial x} = 0.
\end{aligned}
\tag{7.67}
$$

The solution of Eq. (7.62) with conditions (7.67) has the form

$$
\frac{C}{C_0} = \frac{h}{l} + \frac{2}{\pi} \sum_{n=1}^{\infty} \frac{1}{n} \exp\left[-\left(\frac{n\pi}{l}\right)^2 \frac{Dt}{1+K} \right] \sin \frac{n\pi h}{l} \cos \frac{n\pi(l-x)}{l}.
\tag{7.68}
$$

The method of computation by Eq. (7.68) for any value of h/l has been described in a number of papers [20].

Equations (7.64), (7.65), (7.66), and (7.68) may be used for treating the results of geochemical surveys for secondary dissemination aureoles of ore deposits if the distance from the deposit to ground surface is much less than the lateral extent of the deposit. In this case, we observe one-dimensional distribution of the substance diffusing from the deposit, along the vertical from the ground surface, and this agrees with the solved problems.

Diffusion from a Steady Linear Source. If the distance from the surface of the earth to the ore deposit is large in comparison with the transverse width of the deposit, but the length is much greater than its width, the deposit (a vein) may be considered a linear source of the diffusing substance. Let a unit length of the source, parallel to the z axis and passing through point (x', y'), pass $C_0' = \text{const}$ substance into the rock per unit time. The solution of Eq. (7.62) with conditions that no flow of substance into the atmosphere (x = 0) take place and that no diffusing substance is present in the rock at initial time is found by the method of sources and sinks [18]. The solution of the equation has the form

$$
C = -\frac{C_0'}{4\pi D}(1+K)\left[E_i\left\{ -\frac{r^2(1+K)}{4Dt} \right\} + E_i\left\{ -\frac{r_1^2(1+K)}{4Dt} \right\} \right],
\tag{7.69}
$$

where

$$
-E_i(-x) = \int_x^{\infty} \frac{e^{-u}}{u}\, du
\tag{7.70}
$$

is the integral of the exponential function [21], and

$$r^2 = (x - x')^2 + (y - y')^2 \quad \text{and} \quad r_1^2 = (x \div x')^2 + (y - y')^2$$

represent the distances from the point (x', y') to the linear source and to its vertical reflection relative to the earth's surface, respectively.

Numerical calculations according to (7.69) may be made after using tables of the integral exponential function [21]. For the particular case of samples collected at the earth's surface, x = 0 and r = r_1, and in place of Eq. (7.69) we have

$$C = -\frac{C_0'}{2\pi D}(1 \div K) E_i \left\{ -\frac{r^2(1 \div K)}{4Dt} \right\}, \tag{7.71}$$

where

$$r^2 = x'^2 + (y - y')^2.$$

For large time values, by using the first two members of the expansion of function $E_i(-x)$ into a series according to x, we obtain a simpler solution:

$$C = \frac{C_0'}{4\pi D}(1 \div K)\left[\ln \frac{4Dt}{(1 \div K)\, rr_1} - \alpha\right], \tag{7.72}$$

where $\alpha = 0.5772 \ldots$ is Euler's constant.

Equations (7.69), (7.71), and (7.72) may be used in a number of cases for treating results of geochemical surveys for dissemination aureoles, such as when the ore body is a rather deep vein.

Diffusion from a Steady Point Source. If the linear dimensions of an ore body are small as compared with the distance from the earth's surface, the deposit may be considered a point source of diffusing substance. If the deposit lies at point (x', y', z') and from it an amount of substance $C_0' = \text{const}$ passes into the rock per unit time, the solution of Eq. (7.62) for the same initial and boundary conditions as in the problem above for diffusion from a linear source, is found by the method of sources and sinks [18]. The solution has the form

$$C = \frac{C_0'}{\left(4\pi \dfrac{Dt}{1+K}\right)^{3/2}} \int\limits_{0}^{+\infty}\int\limits_{-\infty}^{+\infty}\int\limits_{-\infty}^{+\infty} \left[e^{-\frac{r^2(1+K)}{4Dt}} + e^{-\frac{r_1^2(1+K)}{4Dt}}\right] dx'\, dy'\, dz', \tag{7.73}$$

where

$$r^2 = (x - x')^2 + (y - y')^2 + (z - z')^2;$$
$$r_1^2 = (x + x')^2 + (y - y')^2 + (z - z')^2.$$

Solutions (7.69) and (7.71)-(7.73) may be generalized when the intensity of the sources of diffusing substance changes with time [18].

Thus, the evolved theory of diffusion migration of a substance accounts for reversible adsorption of the diffusing substance and also for ion exchange with the surrounding rocks. Of those processes not accounted for by the theory, the most significant has to do with chemical reactions between the migrating substances and the rocks. If these processes are taken into account, then, the distance of migration of the substances of the deposit will clearly be less

TABLE 7. Time Required for Establishing
an Almost Stationary Flow in Surface Strata.

D, cm²/sec	Time in millions of years for thickness of covering strata, m		
	1000	2000	3000
$5 \cdot 10^{-5}$	2.44	9.76	22
$1 \cdot 10^{-5}$	12.2	48.8	110
$5 \cdot 10^{-6}$	24.4	97.6	220
$1 \cdot 10^{-6}$	122	488	1 100
$5 \cdot 10^{-7}$	244	976	2 200
$1 \cdot 10^{-7}$	1 220	4 880	11 000
$5 \cdot 10^{-8}$	2 440	9 760	22 000
$1 \cdot 10^{-8}$	12 200	48 800	110 000
$5 \cdot 10^{-9}$	24 400	97 600	220 000

than the distance computed above. We should also note that the theory does not consider the evaporation of water at the earth's surface. This effect, leading to a large concentration of diffusing substances at the surface and precipitated there in sediment, needs special investigation.

3. Secondary Dissemination Aureoles of Gas Deposits. Questions on the theory of the formation of secondary aureoles about gas deposits have been examined by Antonov [3, 22], particularly for the purpose of establishing a foundation for geochemical methods of prospecting for oil and gas (gas survey). Antonov examined the basic case in which a steady state is established for a diffusion current of hydrocarbon gas in the surface layers of rock (Chap. 3). The time required for establishing the steady state was computed by Antonov for one-dimensional diffusion of gas, and the results are given in Table 7.

The distribution of stationary concentrations of gas above a deposit is found by solving Eq. (2.9) of stationary diffusion with certain boundary conditions. The characteristic boundary condition for the problem of gas diffusion in the earth's crust is the equation of zero concentration of gas at the earth's surface:

$$x = 0, \quad C = 0. \tag{7.74}$$

Let us consider the solution of Eq. (2.9) with conditions (7.74) for a number of cases having practical interest [3, 22].

Diffusion from a Steady Point Source. We shall set the x axis perpendicular to the surface and the y and z axes in the plane of the earth's surface, so that the coordinates of the gas source are x = H, y = z = 0. We shall assume that the lateral dimensions of the source are so small, as compared with the depth of occurrence, that the projected area on the earth's surface is infinitesimally small, equal to ds · dl. The solution of the problem may be written in the form [22]

$$C(x,r) = \frac{A\, dl\, dS}{h} \sum_{n=0}^{\infty} (-1)^n \left\{ \frac{1}{\sqrt{\frac{r^2}{H^2} + \left[(2n+1) - \frac{x}{H}\right]^2}} - \frac{1}{\sqrt{\frac{r^2}{H^2} + \left[(2n-1) + \frac{x}{H}\right]^2}} \right\}, \tag{7.75}$$

where $r = \sqrt{y^2 + z^2}$ is the polar coordinate of points in the medium, and A is a factor depending on the diffusion coefficient of the investigated gas in the rocks and the pressure under which the gas is found at the source.

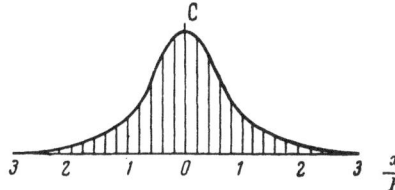

Fig. 42. Distribution of gas concentrations above a point source.

If x ≪ H (with distribution of gas concentrations near the earth's surface), in place of Eq. (7.75) we shall have

$$C(x, r) = \frac{A\,dS\,dl}{H^2} \sum_{n=0}^{\infty} (-1)^n \frac{(2n+1)}{\frac{r^2}{H^2}+(2n+1)^2}.$$ (7.76)

The series (7.76) converges, so that computations from this equation present no difficulties. The distribution of gas concentrations above a point source is shown graphically in Fig. 42.

Diffusion from a Steady Linear Source. Let a linear source of gas of length $2l$ and width dS be oriented along the z axis and lie at a depth H. The distribution of gas concentration in the rock may be represented in the form [22]

$$C(x, y) = \frac{Ax}{H}\,dS \sum_{n=0}^{\infty} (-1)^n \frac{2\frac{l}{H}(2n+1)}{\left[\frac{y^2}{H^2}+(2n+1)^2\right]\left[\frac{y^2}{H^2}+\frac{l^2}{H^2}+(2n+1)^2\right]}.$$ (7.77)

The series (7.77) converges. If $l/h \gg 1$ (in practice $l/h \gtrless (2-3)$), then, in place of Eq. (7.77) we shall have

$$C(x, y) = \frac{Ax}{H}\,dS \sum_{n=0}^{\infty} (-1)^n \frac{(2n+1)}{\frac{y^2}{H^2}+(2n+1)^2} = \frac{Ax}{H} \frac{\pi}{2} \frac{dS}{\cosh\left(\frac{\pi}{2}\frac{y}{H}\right)}.$$ (7.78)

Diffusion from an Infinite Deposit in the Presence of a Linear Diffusion Current. The concentration distribution is found by solving Eq. (3.10) of one-dimensional diffusion with the boundary condition (7.74); as readily seen, it has the form

$$C(x) = Ax.$$ (7.79)

Antonov also examined the problem of stationary diffusion of gas from beds of different configurations (symmetrically and asymmetrically convex and inclined). He showed (see Chap. 2) that the solutions obtained above may be generalized in the case of gas diffusion through a group of beds having different diffusion permeabilities if in these solutions we replace the actual depth H by the effective depth, determined by means of Eq. (2.69).

Solutions (7.75)-(7.79) describe the steady state of a gas current in which the concentration of substance at any point does not change with time. However, the time for establishing the steady state is large (see Table 7), so that a steady state from the diffusion of gas from a deep-seated deposit (H ≈ 2 km, if D ≈ 5·10⁻⁶ cm²/sec) cannot be attained. Consequently, Eqs. (7.75)-(7.79) do not describe gas anomalies above deep-seated deposits. To describe these anomalies it is necessary to solve Eq. (2.6) of nonstationary diffusion with definite initial and boundary conditions. These solutions, for a number of cases, were obtained earlier [Eqs. (7.39), (7.40), (7.43), and (7.44)-(7.48) for a homogeneous medium and (2.76), (2.77), (2.79), and (2.80)

TABLE 8. Distance of Diffusion Penetration of Gases from a Deposit

D, cm^2/sec	Distance of diffusion, m, for duration of migration, millions of years						
	10	25	50	100	200	300	400
$5 \cdot 10^{-5}$	8400	13 300	18 800	26 700	37 700	46 200	53 300
$1 \cdot 10^{-5}$	3750	5 950	8 400	11 900	16 800	20 600	23 800
$5 \cdot 10^{-6}$	2680	4 200	5 940	8 430	11 850	14 500	16 800
$1 \cdot 10^{-6}$	1200	1 880	2 650	3 760	5 300	6 500	7 520
$5 \cdot 10^{-7}$	840	1 330	1 880	2 670	3 770	4 620	5 330
$1 \cdot 10^{-7}$	375	595	840	1 190	1 680	2 060	2 380
$5 \cdot 10^{-8}$	268	422	594	843	1 185	1 450	1 680
$1 \cdot 10^{-8}$	120	188	265	376	530	650	752
$5 \cdot 10^{-9}$	84	133	188	267	377	462	585
$1 \cdot 10^{-9}$	38	60	84	119	168	206	238

for an inhomogeneous medium] and might be used for computing anomalous distributions of gas above deep-seated deposits. The depths of occurrence accessible to geochemical study, as follows from the discussions in the preceding chapters, are determined by the sensitivity of the method of analyzing for the gas. The distance of diffusion migration, computed by Antonov with consideration of the sensitivity of modern instruments and techniques of gas analysis, is shown in Table 8 for different coefficients of gas diffusion in the rocks.

The views we have proposed concerning the diffusion of gases are probably of limited application in prospecting for oil deposits for various reasons (origin, hydrocarbon gases in sedimentary rocks without a deposit being present, low concentrations of gases in the near-surface strata, and so forth).

§ 56. The Question of Oil Formation

The view we have developed in the preceding chapters may be used for describing the formation of oil. This follows from the fact that the formation of oil (as well as minerals) may be represented as a heterogeneous process taking place during migration of dissolved substances, leading to the formation of a new phase (oil).

Let us consider the data from the literature touching on the synthesis of water-insoluble organic substances from water-soluble, which have some relation to the problem of the origin of oil. The possibility of such synthesis was indicated in the experiments of Veber [23] and G. P. Kolpenskii. In Veber's experiments, clay, sand, or silt with low content of organic matter was placed on the bottom of a glass cylinder filled with water. Algae previously subjected to bacterial decomposition were placed in the water above the sediment. Over a long period of time, measured in months, a gradual darkening of the sediments was observed. Analysis of benzene extracts from the sediments showed that the amount of water-insoluble substances in the sediment increased substantially during the experiments. One possible explanation of the observed phenomenon is that the organic matter, formed by decomposition of the algae, diffused into the sediment. It is possible that this organic matter was water-soluble (since diffusion in this case could take place at a noticeable rate). It would then form in the sediment by chemical reaction between the decomposition products of the algae.

The possibility of synthesizing water-insoluble organic substances from water-soluble has been confirmed by experiments of G. P. Kolpenskii, a coworker at the All-Union Scientific Research Petroleum-Geological Exploration Institute (VNIIGNI), completed in 1957. In his experiments, Kolpenskii collected several tens of samples of bituminous rocks having different lithic composition. The rock material consisted of rock chunks having a volume of 10-15 cm^3.

By means of prolonged extraction in organic solvents, the samples were completely deprived of their bitumens; they did not fluoresce in ultraviolet light. The samples were than placed in extractors, in an atmosphere of light hydrocarbons (methane and ethane). After 4-6 months it was found that all the samples again contained bituminous substance and fluoresced under ultraviolet light. The sources of the synthesis of bituminous substances might have been either hydrocarbons (methane, ethane) making up the atmosphere in the extractor, or water-soluble organic substances that may have been washed out of the pores by the organic solvents in the process of extraction.

Since the first source is excluded from theoretical considerations, the water-insoluble heavy hydrocarbons were synthesized from water-soluble organic substances contained in the rock. The products of reaction of the synthesis that formed in the time preceding the synthesis were removed by extraction; as a result the reaction continued. Despite the incompleteness of these investigations, cut off and unpublished because of the untimely death of Kolpenskii, they are of great interest, since they point to the possibility of synthesizing heavy hydrocarbons from water-soluble organic substances. The presence of water-soluble organic matter in rocks is well known, and we shall not dwell longer on this subject.

The synthesis and accumulation of heavy hydrocarbons in rocks is part of the problem concerning the origin of oil. At the present time, some geologists distinguish a theory of petroleum source rocks as the sources of the initial material of the oil, by which they mean water-soluble hydrocarbons disseminated through the source rocks. It is known that insoluble hydrocarbons cannot ordinarily migrate from the source rocks to the reservoir rocks where the oil deposit is formed. The hypothesis of oil formation from water-insoluble organic matter (and also the hypothesis of inorganic origin of oil) cannot be explained by the accumulation of oil in reservoir rocks. No known hypotheses can explain either why oil is found in individual droplets in pores or entirely fills the cavities of shells or the pores of corals. It may be pointed out that these features have been observed where the surrounding rocks contain absolutely no oil-like substances.

The indicated difficulties may be partly overcome if it is assumed that oil is formed from water-soluble organic matter contained in the source rocks. We might state that at first no water soluble organic matter is present in a sandy reservoir rock, or that the concentration of such matter is lower than in the surrounding source rocks. Diffusion of the dissolved organic substances to the reservoir rocks then begins. In the reservoir rocks the substances may interact with each other, giving rise to insoluble phases (gas, oil, or other insoluble organic substances). Consequently, an irreversible chemical reaction takes place in the reservoir rock, by which the diminution of dissolved soluble organic matter in the reservoir rock because of the reaction will be compensated by diffusion from the surrounding rocks (Chap. 4). As a result, a steady-state process is established, in which the rate of chemical reaction will be equal to the supply rate of reacting substances to the reservoir rocks by diffusion. Collecting in different parts of the reservoir rocks, large volumes of oil and gas may unite, float upward, and make up the well-known system of gas, oil, and water distributed in the reservoir rock in accordance with their specific gravities [24]. The formation of a deposit will continue to go forward until the dissolved organic matter in the surrounding source rocks is completely gone.

Thus, the theory we have developed concerning heterogeneous processes of migration may be used to explain dissemination processes and accumulation not only of minerals but also of organic substances in the earth's crust.

LITERATURE CITED

1. Solovov, A. N., Principles of the Theory and Practice of Metallometric Surveys [in Russian], Izd. AN Kaz. SSR, Alma-Ata (1953).
2. Saukov, A. A., Geochemical Methods of Prospecting for Mineral Deposits [in Russian], Izd. MGU (1963).
3. Antonov, P. L., in: Direct Methods of Prospecting for Oil and Gas [in Russian], p. 5 (1964).
4. Dubov, R. I., in: Selected Lectures on Geochemical Methods of Prospecting for Ore Deposits [in Russian], Izd. AN Kaz. SSR, Alma-Ata (1963).
5. Dubov, R. I., in: The Geochemistry of Ore Deposits [in Russian], Izd. Nauka, Moscow, p. 117 (1964).
6. Panchenkov, G. M., and Lebedev, V. P., Chemical Kinetics and Catalysis [in Russian], Izd. MGU (1961).
7. Panchenkov, G. M., Zh. Fiz. Khim., Vol. 38, No. 3 (1964).
8. Korzhinskii, D. Z., Vserossiisk. Mineral. Obshch., Vol. 71, p. 750 (1942).
9. Bosthoud, A., J. Chim. Phys., Vol. 10, p. 624 (1912).
10. Solovov, A. P., Geologiya Rudnykh Mestorozhdenii, Vol. 8, No. 3 (1966).
11. Pospelov, G. L., Izv. Sib. Otd. AN, Geologiya i Geofizika, No. 11, p. 28 (1962); No. 11, p. 40 (1962); No. 10, p. 20 (1963).
12. Pospelov, G. L., and Kaushanskaya, P. I., Kolloid. Zh., Vol. 25, p. 215 (1963).
13. Pospelov, G. L., and Kaushanskaya, P. I., Izv. Sib. Otd. AN, Geologiya i Geofizika, No. 9, p. 41 (1962); No. 5, p. 35 (1965).
14. Afanas'ev, P. B., Zel'dovich, Ya. B., and Todes, O. M., Zh. Fiz. Khim., Vol. 23 p. 1965 (1949).
15. Zel'dovich, Ya. B., and Todes, O. M., Zh. Fiz. Khim., Vol. 23, p. 180 (1949).
16. Gerasimov, Ya. I., and others, A Course in Physical Chemistry [in Russian], Goskhimizdat (1963).
17. Gnedenko, B. V., A Course in Probability Theory [in Russian] (1954).
18. Carslaw, H. S., and Jaeger, J. C., Conduction of Heat in Solids, Clarendon Press, Oxford (1947).
19. Van der Pol, B., and Bremmer, H., Operational Calculus based on the Two-sided Laplace Integral, Cambridge University Press (1955).
20. Barrer, R. M., Diffusion in and through Solids, Cambridge University Press (1941).
21. Yalmke, E., Emde, F., and Lesch, F., Special Functions [in Russian] (1964).
22. Antonov, P. L., Neftyanoe Khozyaistvo, No. 5, p. 20 (1934).
23. Veber, V. V., in: Accumulation and Transformation of Organic Matter in Recent Marine Sediments [in Russian], p. 37 (1955).
24. Sokolov, V. A., Essays on the Origin of Oil [in Russian], Gostoptekhizdat, Moscow-Leningrad (1948).